Study Guide to Accompany

ESSENTIALS OF THE DYNAMIC UNIVERSE
AN INTRODUCTION TO ASTRONOMY

THIRD EDITION

Theodore P. Snow
University of Colorado

Prepared by
Jeffrey O. Bennett
University of Colorado

Catharine D. Garmany
University of Colorado

Theodore P. Snow
University of Colorado

West Publishing Company
St. Paul   New York   Los Angeles   San Francisco

COPYRIGHT © 1990 by WEST PUBLISHING CO.
50 W. Kellogg Boulevard
P.O. Box 64526
St. Paul, MN 55164-1003

97 96 95 94 93 92 91 90   8 7 6 5 4 3 2 1 0

ISBN 0-314-70766-2

Contents

iii

Preface

This Study Guide is designed to accompany the textbook *Essentials of the Dynamic Universe, Third Edition*, by Theodore P. Snow. Each of the chapters in the text has a corresponding chapter in this guide.

The Study Guide will help you to understand the more difficult concepts from the text. Each chapter begins with a brief overview highlighting some of the major points made in the textbook. Following the overview, each chapter contains a brief list of new words and phrases which were introduced in the text. You should review the list and return to the text for an explanation of any words or phrases that you do not understand.

Many of the chapters contain one or more *Exercises* which are designed to help clarify important concepts from the text. Some of the Exercises are answered at the end of the chapters. Each chapter also contains a multiple choice Self-Test, with answers at the end. In reviewing your answers, be sure you understand not only the correct answer, but also why the other choices are incorrect.

Finally, following the Self-Test, each chapter offers solutions to selected Review Questions from the textbook. In general, solutions are given to the more difficult questions. Those questions that simply require careful reading of the text are not answered for you.

To summarize, we recommend the following strategy for study:
1) Read the chapter in your textbook.
2) Read the overview and identify the Words and Phrases in the corresponding chapter of this Study Guide; if any of the material seems unfamiliar you should go back and review the textbook.
3) If there are any Exercises in this Study Guide for the chapter you are studying then work through them.
4) Check your understanding of the material by working through the Self-Test in this Study Guide.

5) Further test yourself by answering the Review Questions from the textbook; solutions to some of these are presented in this Study Guide.

6) If you wish to study a topic further, consult the list of Additional Readings given in each chapter of the textbook.

Many have contributed to the development of earlier editions of this Study Guide. We particularly wish to acknowledge C. M. Snow, K. S. Bjorkman, S. J. Shawl, and D. Simon for their help. We are grateful to Terry Armitage for proofreading, numerous suggestions, and production of the camera-ready text for this third edition.

We hope that this Study Guide will be a valuable resource for students taking an introductory course in astronomy, as well as for those who already are somewhat familiar with the subject. We all can appreciate the visual beauty of astronomy; this guide is designed to help those of you who wish to appreciate the intellectual beauty as well.

Jeffrey O. Bennett

Catharine D. Garmany

Theodore P. Snow

October, 1989

1
The Essence of Astronomy

The subject of astronomy encompasses the entire universe, from the smallest scales to the largest. Modern astronomy is more properly called *astrophysics* since we now are capable of answering fundamental physical questions about the ways in which planets, stars, and our universe work. Astronomy is one of the oldest areas of human inquiry, because it allows us to ask fundamental questions about who we are, where we come from, and how our universe works.

In ancient times the interpretation of what was seen in the sky often was based on myth. During the classical period of ancient Greece astronomy became more quantitative, but still held on to mistaken views such as the belief that the Earth is the center of the Universe. The observations and beliefs of that time became the psuedoscience of astrology in which *apparent* motions of heavenly bodies, such as those of the planets among the stars, are believed to directly influence human affairs. The inherent fallacy of astrology is easily understood by the simple realization that planets *do not* move among the stars at all! Rather, the planets orbit the Sun, just as does the Earth, while the stars are thousands of times further distant. The motions charted by astrologers are simply illusions created by the fact that our own Earth is a moving vantage point. We note that observers on other planets, on the Moon, or anywhere else in the Universe would see different apparent motions.

In the seventeenth century astronomers began to apply the laws of physics in order to try and *understand* their observations of the heavens. It was at this time that astrology became permanently disconnected from astronomy, since astronomy became a *science* while astrology remained superstition.

Sizes and distances in our universe are often incomprehensible in magnitude. Nevertheless, it can be useful to imagine scale models in order to get some feel for astronomical numbers. For example, imagine that our Sun is the size of a grapefruit. Then the largest planet — Jupiter — would be about the size of a marble and located about 75 meters away; Pluto, the smallest planet, would be about the size of the period at the end of this sentence and located, on average, about half a kilometer from the grapefruit–sized Sun. The Earth would be about the size of a ball–point from a pen, located about 15 meters from the Sun. Indeed, by visualizing the Sun as grapefruit–sized you could fit a model solar system with all the other planets on most college campuses. And yet, on that scale, one would have to travel some 4000 kilometers (2500 miles) just to reach the nearest other stars (the triple star system of Alpha Centauri)!

Since there are some 100 billion to 1 trillion stars in our Milky Way galaxy, we would need a greatly expanded scale to visualize the galaxy. If we imagine our entire solar system to fit in a circle 8 centimeters across, then the nearest star would be about 300 meters away, and our galaxy about the size of the United States. To imagine the extent of the observable universe, one must remember next that our galaxy is only one of ten to one hundred *billion* galaxies!

Distances in astronomy are sometimes measured in terms of the distance travelled by light in a given period of time; thus we speak in terms of *light–years*. (Note that a light–year is a unit of distance and *not* of time!) For example, if a star is ten light–years away, it means that light from that star requires ten years to reach us. Anything else, like a spaceship, would of course take much longer to travel that same distance. Because light takes time to reach us from its source, we see distant objects as they appeared at sometime in the past when the light left on its way. Thus, by looking out to great distances in the universe we also are looking back into time.

Words and Phrases to Understand

astrology	light year
astronomical unit	Magellanic Clouds
astronomy	meteor
astrophysics	Milky Way galaxy
celestial sphere	nebula
cluster of stars	planets
cluster of galaxies	revolution
comets	rotation
ecliptic	supercluster
interstellar gas	

Exercise 1: What Do Scientists Do?

It is sometimes said that the best scientist is the one who asks the right questions. As Stephen Jay Gould writes in *Hen's Teeth and Horse's Toes* (Norton & Co.: New York 1984), the prerequisite for great science is presenting a question in a new and different way. While very few people have the insight to ask questions which will change the direction of scientific research, anyone who studies science soon realizes that science is a method of dealing with problems, not a collection of facts. Astronomy is no exception. Although it differs from the laboratory sciences in that few aspects of astronomy can be directly examined, it shares with other branches of science a method in which the development of general ideas leads to specific theories which can be tested and which must comply with the known laws of physics. Since astronomy covers such vast territory, organization of the available data is especially important, and it is often this organization which leads someone to ask a new and crucial question.

Go out on a clear night and look at the sky for awhile. After you come in, make a list of the questions that came to mind while you were outside. These would be questions about the things you saw, perhaps about their sizes and distances, their motions, their compositions, or their relationships to each other. Having the same observational tools (your eyes) as our ancient ancestors, you will be able to ask the same questions they did. During your study of astronomy, you will see how the ancients eventually found answers to their questions, and you will see how many of your questions are answered. As your knowledge increases, you will be able to ask more sophisticated questions, and will come to understand some of the important ones being asked today by astronomers at the forefront of research.

3

Exercise 2: The Dual Meaning of "Milky Way"

The term *Milky Way* has a dual meaning. Historically, it refers to the bright, milky–colored band which we see stretching across our sky on clear, dark nights. Today, we also use the term as the name of our own spiral galaxy. In this exercise, we will try to understand the relationship between the two meanings of the term *Milky Way*.

At different times, and from different latitudes, we see different portions of the Milky Way band in our sky. Nevertheless, if you study a star map, you will see that the Milky Way band stretches a full 360° around the sky. The band is what we see when we look at the millions of stars, plus gas and dust, along a line–of–sight into the disk of our galaxy, in any direction. When we look away from the disk of the galaxy (the rest of the sky), we see far fewer stars because the disk is so thin in comparison with its diameter. Further, when we look away from the disk, we have a clear view to objects located outside of our own galaxy. Note that the Milky Way band is wider in the direction of the center of the galaxy (towards Sagittarius).

To fully understand the dual meaning of *Milky Way*, imagine that our galaxy is a pancake (bulging in the center). Our solar system is located in the pancake about 2/3 of the way from the center towards the outer edge. Use this analogy to explain why we see the Milky Way band in our sky.

Exercise 3: Large Numbers

We deal with large numbers in astronomy. While these large numbers may be easy to say or write, they often are truly incomprehensible in magnitude. Nevertheless, analogies often can impress us with at least some understanding of the size scales in the Universe. For example, a typical spiral galaxy contains something in the neighborhood of 100 billion (10^{11}) stars. Suppose that you decide to make an accurate count of all 100 billion stars in some galaxy (and neglect the fact that many are too dim to see, or may be hidden from view by gas and dust). If you could count them at a rate of one star each second, how long would it take you to finish? (Obviously, it will take 100 billion seconds. Convert this to years.)

Our own Milky Way may have as many as a trillion stars. How long would it take to count to one trillion (10^{12})?

Self Test

1. Today, astronomy is synonymous with
    a. astrology
    b. astrophysics
    c. space exploration
    d. geophysics.

2. Which of the following descriptions best fits our solar system?
    a. A group of several stars, with an overall diameter of about 10 light–years
    b. A central star surrounded by 9 planets of approximately equal size
    c. A central star containing 99% of the system's mass, surrounded by 9 planets plus numerous moons, asteroids, comets and other debris
    d. A large, swirling collection of stars about 100,000 light–years in diameter
    e. Everything in the known universe.

3. The particular stars that can be seen on any given night depend on
    a. the brightness of the Moon
    b. the season of the year
    c. the time of night
    d. our geographic latitude
    e. all of the above.

4. Which of the following correctly goes from the smallest scale to the largest?
    a. Earth, galaxy, solar system, cluster, supercluster, universe
    b. Earth, solar system, universe, cluster, galaxy, supercluster
    c. cluster, Earth, universe, solar system, galaxy, supercluster
    d. Earth, solar system, galaxy, cluster, supercluster, universe.

5. An astronomical unit is
    a. a distance equal to the average Earth–Moon distance
    b. a distance equal to the average Sun–Earth distance
    c. the distance light travels in one year
    d. one course in elementary astronomy.

6. A light–year is
    a. a measure of time
    b. a measure of distance
    c. the velocity of light
    d. the distance from the Earth to the Sun.

5

7. The path on which the Sun appears to move among the stars around the sky each year is called
   a. the Milky Way
   b. the ecliptic
   c. the equator
   d. the meridian
   e. the equinox.

8. Suppose we imagine the Sun to be about the size of a grapefruit. About how far away would the *nearest stars* (the three stars of alpha centauri) be found on the same scale?
   a. 10 yards
   b. 100 yards
   c. 250 miles
   d. 2500 miles.

9. About how many stars are in a typical spiral galaxy?
   a. 100,000
   b. 100 million
   c. 1 billion
   d. 100 billion.

10. The diameter of a typical spiral galaxy is about:
    a. 100,000 miles
    b. 100,000 A.U.
    c. 1000 light–years
    d. 100,000 light–years.

Answers to Selected Review Questions from the Text

4. The fact that all of the planets follow nearly the same strip through the sky (near the ecliptic) tells us that the orbits of all the planets lie in approximately the same plane. This, in turn, provides evidence that the planets formed from a disk–shaped cloud of material.

6. Only on the equator, where the north and south poles lie on the opposite horizons, can all the stars be seen over the course of one year. When you are in one hemisphere, a fraction of the stars in the opposite hemisphere are forever below the horizon, and this fraction increases as you travel toward the pole. In order to observe all parts of the sky equally, it is necessary to have observatories in both hemispheres.

7. The Sun's angular diameter is about 30 arcminutes. Since there are 60 arcminutes in one degree, this is the same as 1/2 degree. Since there are 60 arcseconds in each arcminute, it also is the same as $60 \times 30 = 1,800$ arcseconds.

10. If galaxies formed 10 billion years ago, then we must look more than 10 billion light–years distant to see the universe at a time before galaxies existed.

Answer to Exercise 3

It would take over 3,000 years (about 3,200) to count to 100 billion at a rate of one per second.

Answers to Self–Test

1–b, 2–c, 3–e, 4–d, 5–b, 6–b, 7–b, 8–d, 9–d, 10–d.

2
Cycles and Seasons: Motions in the Sky

Humanity was mystified by apparent celestial motions for thousands of years because of a belief that the Earth was the center of the universe. Once it was recognized that the Earth orbits the Sun once each year and that the other planets orbit the Sun with periods that are determined by their distance, then the mystery of celestial motion was solved.

The largest apparent motion is due to the rotation of the Earth every 24 hours. (Note: the true rotation period of the Earth is called one *sidereal* day and is approximately 23 hours, 56 minutes in length.) Since the Earth rotates from west to east, objects in our sky — including the Sun, Moon, planets, and stars — *appear* to rise from the east and set in the west. This daily motion also is the basis of our timekeeping system.

The second obvious motion in the sky is that of the Moon, which orbits the Earth every 27 days. The Moon appears to move eastward *relative* to the stars; that is, if the Moon rises in the east at the same time as some star, then, since the Moon will move slightly eastward from the star over the course of the night, the Moon will set in the west slightly later than the star. This motion of the Moon relative to stars can be noticed over a period of just a few hours. Night–by–night, this motion leads to the changing phases of the Moon.

The Sun also appears to move eastward *relative* to the stars as a result of the Earth's orbital motion around the Sun once each year. Because of this motion, stars appear to set about 4 minutes earlier each night, and we see different constellations in the evening sky at different times of year. Thus, we can speak of the "autumn sky" or the "summer sky" (etc.) in reference to the constellations that we see in those seasons.

actually rotates counter clockwise

8

The orbital plane of the Earth around the Sun is called the *ecliptic*. The Earth's axis of rotation is tilted by 23.5° to the ecliptic, so that the inclination of the Sun in our sky changes over the course of the year; this is the cause of our seasons. The orbits of the Moon and all of the planets lie nearly, but not exactly, in the ecliptic; the narrow path in the sky through which the Moon, Sun, and planets can be found is called the *zodiac*. *If* the orbit of the Moon were exactly aligned with the ecliptic, then we would see a lunar eclipse at every full moon and a solar eclipse at every new moon. Since the orbits of the Moon and Earth are not perfectly aligned, however, eclipses are much rarer and can occur only when the Moon passes through the Earth's orbital plane at the same time that the phase is either new (solar eclipse) or full (lunar eclipse).

The planets change their apparent positions among the stars both because of their own motions around the Sun, and because of the Earth's motion around the Sun. Mercury and Venus, since they are closer to the Sun than the Earth, swing from one side of the Sun to the other and thus can be seen only around the times of Sunset or Sunrise (i.e., never at midnight). The outer planets appear to move gradually eastward relative to the stars, their rate of motion depending on the size of their orbit. Jupiter, for example, takes nearly 12 years to sweep through all of the zodiac constellations on its eastward track. Although the outer planets generally move eastward relative to the stars, each appears to move westward for a short time each year as the Earth passes by in its orbit. This apparent backwards, or *retrograde*, motion mystified ancient astronomers, but in fact is simply explained once we understand that the Earth is not the center of the universe.

The dawn of understanding about the sky began to occur in many parts of the world around 500 B.C. Although parallel developments took place in Asia, India, the Americas, and the Mediterranean region, it is from the latter that we have inherited most of our nomenclature and ideas. Babylonian astronomy dates back to as far as 1700 B.C., and cuneiform tablets from that period show that these people charted many of the cyclic motions of the sky. By 200 B.C., people of this region were able to predict eclipses and planetary motion, and they developed a cosmology to fit their world view. The Greeks borrowed from the Babylonians, but went much further. The idea that natural events can be understood, and described mathematically, originated with the teachings of Pythagoras. The realization that the Moon shines by reflected Sunlight, and thus the correct understanding of eclipses, dates from this period.

Plato's teachings, that understanding should come primarily from reason rather than direct observation, dominated Western thinking for nearly 2000 years. One of his students was Aristotle, who was the first to adopt physical laws to explain the universe. With the development of geometry, it became possible for the Greeks to deduce such things as the relative sizes and distances of the Moon and Sun, and the size of the Earth. Perhaps the greatest Greek astronomer was Hipparchus, who measured star positions, discovered precession of the stars, refined the methods of determining sizes through the use of geometry, and developed the magnitude system still in use by astronomers today. Greek astronomy ended in about 150 A.D. with the work of Ptolemy, who compiled an encyclopedia of all astronomical knowledge.

Names

Aristarchus	Plato
Aristotle	Ptolemy
Eratosthenes	Pythagoras
Hipparchus	

Words And Phrases

angular diameter
Arctic Circle
Antarctic Circle
celestial equator
conjunction
constellations
declination
deferent
diurnal motion *daily*
eclipses, solar and lunar
ecliptic
epicycle
equinoxes, vernal and autumnal
Gregorian Calendar
heliocentric
inferior planets
Julian Calendar
meridian
Milky Way

opposition
orbital motion
phases of the Moon
planets
precession
quadrature
right ascension
retrograde motion
rotation
sidereal period
solar day
solstice, summer and winter
stellar parallax
synchronous rotation
synodic period
superior planets
Tropics, Cancer and Capricorn
zenith
zodiac

Exercise 1: Spatial Orientation

In astronomy it is very important to be able to visualize things spatially; that is, to mentally view something from different directions when it is impossible to actually do so. The following exercise is designed to help you develop these skills.

11

We start with our direct observations: on some particular clear night in spring we go out to stargaze. We know the cardinal directions: north, south, east and west. The first quarter moon is on our meridian just as the Sun sets, and someone asks what the Moon looks like at that moment to people several time zones to the east (for example, if we are in Colorado, how does the Moon appear to a New Yorker?) To answer this, let us imagine that we could view the Earth–Moon system from a point in space far above the north pole:

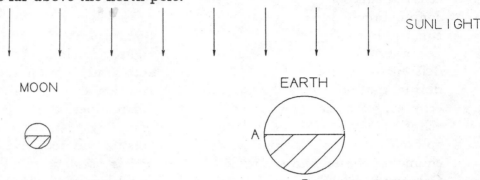

Our location at Sunset is at point A. About 6 hours later the Earth's rotation has carried us to point B, but the Moon's motion is negligible. Therefore the Moon is setting in the west. Keep in mind that the direction "east" is simply the direction in which the Earth is rotating, and "west" is the opposite direction.

Now ask yourself the following questions: what is the phase of the Moon when it is overhead at midnight? During what hours can you see the third quarter moon? In general, what fraction of the time is the Moon in the daytime sky? What does the Moon look like to people several time zones to the east, our original question?

Now let's expand the problem. The time is about 2 hours after Sunset. Close to the western horizon we see Venus. Mars is on our meridian, and Jupiter has just risen. What would the solar system look like if we could view it from far above the ecliptic plane? We would see the following, with the orbital paths of the planets partially drawn for clarity.

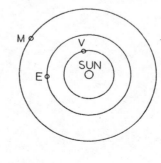

Try the following problems in the same way (your answers will not be exact, but that is not important):

1. You get up an hour before Sunrise. The Moon is just setting. Just above the eastern horizon you identify Mercury, and higher in the sky, Venus. Jupiter is about to set and Saturn is on your meridian. Draw the planets as they would appear from above the ecliptic plane.

2. With some information about the rate at which the planets move, you can predict what the sky will look like at some time in the future. Skip ahead in the text and find out how much of its orbit each planet will complete in three months. Now redraw your figure from above, not forgetting that the Earth will move also! What will the morning sky look like now?

3. Finally, let's turn the problem around. On March 10, 1982, the configuration of the planets as seen from above the solar system was as shown on the last page. This is as close to mutual conjunction as the planets will get for another 179 years, and it gained attention because of a book entitled "The Jupiter Effect" written in 1974. The authors, John Gribbin and Stephen Plagemon, predicted that serious disasters in the form of Earthquakes would occur on this date. Scientists studying their theory found no basis for it, and even the authors revised their views after the book's publication, but many people heard about it and were apprehensive. Nothing catastrophic occurred on March 10, but many people observed the planetary configuration. From the figure, describe what the planets looked like as seen from Earth and when you could have seen them best.

Exercise 2: Estimating Angular Size

When we measure positions of objects in the sky we talk in terms of *angular* distances since we have no sense of depth perception with which to talk of real distances. Measuring angular distance is very simple in principle: it is just like using a protractor to measure angles on a piece of paper. In practice, however, careful measurement of angular distance requires the use of precision instruments. Nevertheless, we can easily make rough estimates of angular distance simply by using an outstretched fist. In this exercise, you will calibrate your own fist so that you can estimate angular sizes.

1. First, you will need to know your latitude. Look it up. Remember that the North Celestial Pole (NCP) can be found at an altitude above your northern horizon equal to your latitude.

2. Next, go outside on a clear night and find the north star, Polaris. Polaris is the end of the handle of the little dipper (see star charts in your book) and also is known as the north star because it is located less than 1 degree from the NCP. Since you already know your latitude, you also know the approximate altitude of Polaris.

3. Extend your arm towards due north, and clench your fist. Looking through only one eye, align the bottom of your fist with the north horizon. Then, by "stacking" your two fists upward, count the number of your fists which are required before you reach Polaris. How many of your "angular fists" above the horizon is Polaris located?

4. Since you know the altitude of Polaris, you can now calculate the angular size of your outstretched fist. You can also calculate the angular size of a single knuckle, etc. Use your fist to measure angular distances between some objects in the sky. For example, you can measure the noon altitude of the Sun each day in order to watch how it changes with the seasons.

Self Test

1. At the summer solstice (June 21) the declination of the Sun is
   a. $+23\frac{1}{2}°$
   b. $-23\frac{1}{2}°$
   c. $0°$
   d. different every year.

2. If the Moon completely covers the Sun as seen by an observer on Earth, it is called
   a. a total lunar eclipse
   b. a total solar eclipse
   c. a partial lunar eclipse
   d. a partial solar eclipse
   e. an annular eclipse.

3. The first quarter moon would be expected to
   a. rise at noon
   b. set at noon
   c. rise at 6 a.m.
   d. set within 3 hours after Sunset.

4. We experience seasonal variation of temperatures on the Earth primarily because
   a. the orbit around the Sun is not circular, so the amount of heat we receive varies
   b. the axis of rotation is tilted with respect to the orbit
   c. the heat from the Sun varies in a regular way due to Sunspots and solar flares.

5. The apparent motion of the planets with respect to the stars is
   a. visible after a few hours of observing
   b. always eastward
   c. generally eastward, but occasionally westward
   d. not visible to the naked eye.

6. An observer on the Earth's equator will see:
   a. stars move parallel to the horizon as the Earth turns
   b. the entire celestial sphere in the course of a year
   c. the north celestial pole but not the south pole
   d. the Sun pass through the zenith every day.

7. The effect of precession is to
   a. gradually change the inclination of the Earth's axis to the ecliptic plane
   b. gradually change the shape and appearance of the constellations
   c. gradually change the constellations that define the zodiac
   d. gradually shift the intersection of the equator and ecliptic along the ecliptic.

8. The first to assert that the Earth is round was
   a. Pythagoras
   b. Aristotle
   c. Plato
   d. Ptolemy.

9. The cosmology of Plato and Aristotle included
   a. a flat Earth at the center of the Universe
   b. a spherical Earth at the center of the Universe
   c. a spherical Sun at the center of the Universe
   d. a spherical Earth at the center of the solar system.

10. The earliest recorded astronomical data were found in
   a. Stonehenge, England
   b. China
   c. Iraq
   d. Mexico.

Answers to Selected Review Questions from the Text

4. If the Earth's axis were not tilted then we would have no seasons. Days would always be split evenly between daylight and darkness. If the Earth's axis were tilted by 90°, then seasonal variations would be much more extreme.

5. Your astrological sign is not in accord with the true position of the Sun among the constellations because of the effects of precession. Astrological signs were made up over 2000 years ago. Since one precession cycle takes 26,000 years, the Earth's axis has precessed about 1/12 of a cycle since that time. Thus, all of the astrological signs are off by about a month.

7. At new Moon, the Moon is located nearly on our line–of–sight to the Sun. Thus, the Moon rises and sets at about the same time as the Sun when its phase is new. Since the Moon is near the Sun in the sky, it will not be visible during the night.

9. If you doubled the Moon's distance its angular diameter would be halved; if you halved the distance then the angular diameter would double. If the angular diameter were halved (further distance) then all solar eclipses would be *annular* since the Moon could not totally block the Sun. If the angular diameter were increased then total solar eclipses would occur more often and tend to last longer.

10. The motions would appear very different on Venus. Since Venus rotates in the opposite sense to that of the Earth (i.e, Venus rotates from east–to–west), objects would rise west and set east. Also, because Venus rotates very slowly, objects would move through the sky very slowly. Venus has no moon. The motions of the planets through the constellations as seen from Venus would be similar to those seen from the Earth, since the orbit of Venus is quite like the orbit of the Earth except for its smaller radius and (therefore) faster orbital speed.

Answers to Self Test

1–a, 2–b, 3–a, 4–b, 5–c, 6–b, 7–d, 8–a, 9–b, 10–c

3
The Renaissance and the Laws of Motion

With the re–awakening of an intellectual spirit in the fifteenth century, European astronomy progressed substantially from that of the ancient Greeks. The major discoveries of the period can be traced primarily to five individuals: Copernicus, Tycho Brahe, Galileo, Kepler, and Newton.

Nicholas Copernicus, encouraged by the climate of inquiry of his time, revived the idea of a heliocentric (Sun–centered) solar system which had been suggested by the Greek scientist Aristarchus nearly 2,000 years earlier. Because Copernicus retained perfect circles for the shapes of the orbits in his system, his predictions of planetary positions were no more accurate than those of Ptolemy's geocentric (Earth–centered) system. However, the Copernican system provided a much more natural explanation of the apparent retrograde motions of planets, and also allowed him to calculate and to find pleasing regularities in the relative distances and orbital speeds of the planets about the Sun. The work of Copernicus was published in 1543.

In the latter part of the 1500's, Tycho Brahe systematically compiled observations of planetary positions with a degree of accuracy far surpassing anything previous. His observations convinced him that the planets orbited the Sun rather than the Earth; however, the fact that he still was unable to detect parallax of the stars led him to conclude that the Earth was stationary and orbited by the Sun. This strange model of the solar system never caught on with other scientists.

Despite his own lack of success at synthesizing his observations into a coherent theory of the solar system, Tycho Brahe's data provided the basis for the work of Kepler. Using Tycho's data, particularly that concerning Mars, Kepler discovered the correct laws of planetary motion. Kepler's first law states that each planet orbits the Sun in the shape of an ellipse with the Sun at one focus. His second law states that planets sweep out equal areas in equal time intervals as they orbit the Sun; one consequence of this law is that planets move fastest in the part of their orbit near perihelion (the point nearest the Sun) and slowest near aphelion (the point furthest from the Sun). Kepler's third law states that the square of the orbital period of a planet is proportional to the cube of the semimajor axis of its orbit; a consequence of this law is that the closer a planet is to the Sun (on average), the faster its average orbital speed. Using his three laws, Kepler was able to predict planetary positions with an accuracy almost 100 times that of earlier systems. Kepler's first two laws were published in 1609 and his third in 1619.

Galileo Galilei, a contemporary of Kepler, developed much of the basis for modern experimental physics. For astronomy, Galileo built his own telescope (he did not invent it, but built his own after reading about others) and turned it to the sky. With the telescope he made a number of important discoveries including lunar craters and mountains, the satellites of Jupiter, the phases of Venus, and the existence of Sunspots. These discoveries helped to prove the validity of the heliocentric system. They also got him into a great deal of trouble with the Church, which was rather intolerant of Galileo's views.

Sir Isaac Newton certainly ranks as one of the foremost scientists of all time; his work is the basis for much of our present understanding of the mechanics of the heavens. Expanding on the work of Galileo, Newton quantified the concepts of mass (and the related concepts of weight and inertia), force, acceleration, and velocity. His three laws of motion summarize these ideas. He also formulated the law of universal gravitation, becoming the first to understand that the same force which causes objects on the Earth to fall also holds the Moon in orbit of the Earth and the planets in orbit of the Sun.

The law of universal gravitation states that every pair of bodies is attracted by a force proportional to the product of their masses and inversely proportional to the square of the distance between them. With the law of gravitation and his laws of motion, Newton was able to explain and expand upon Kepler's laws of planetary motion. In particular, Newton's modification of Kepler's third law allows us to calculate the mass of any object which is orbited by another. For example, we can calculate the mass of the Sun by knowing the semimajor axis and orbital period of any planet; we can calculate the mass of Jupiter by knowing the semimajor axis and orbital period of one of its moons; we can even use this law to calculate the masses of stars in binary systems.

Newton also was able to explain the origin of tides. Tidal forces are simply the result of the difference in gravitational force across an object. For extended bodies which are relatively close to one another, the gravitational force varies across the diameter of the body. In the case of the Earth, this differential force comes primarily from the Moon, and causes the Earth to be stretched somewhat along the line-of-sight to the Moon. Thus, we see high tides both facing and opposing the Moon at all times, and locations on Earth pass through two high tides each day as the Earth rotates. Tidal friction can cause the rotation of a body to slow over time.

Names

Copernicus	Newton
Tycho Brahe	Galileo
Kepler	

Words and Phrases

acceleration	Kepler's laws
angular momentum	law of gravitation
aphelion	luminosity
center of mass	mass
ellipse	Newton's laws
energy, potential and kinetic	perihelion
escape velocity	power
focus, foci (pl.)	tidal force
force	weight
inertia	

Exercise 1: Learning to Like Equations

Many students, when confronted with an equation, have a tendency to panic. Even if they know which equation applies to the problem at hand, and their algebra is up to the necessary manipulations, they still feel unable to tackle the problem. Let's discuss some of the reasons for this. Many of the physical quantities with which we have been dealing are unfamiliar and difficult to visualize, like force and momentum. Then there is the question of what units to use. Should you use miles or kilometers? Will it make a difference if you use grams or kilograms?

There are several important ideas about equations which will make working with them easier, and allow you to check your work and become more confident. First, an equation such as the law of gravity must be true no matter what units are used to measure the physical quantities. Therefore an equation may be solved using different sets of units. However, it is important to realize that the units must balance. An equation is like a seesaw: the two sides are equal by definition. But unlike the seesaw in the park where you can balance your kid brother and a sack of potatoes, an equation demands the same quantities on both sides. This concept can be very useful in solving problems and in checking your solution. If you find that you have set kid brothers equal to potatoes, something is wrong.

As an example, consider the relation $F = m \times a$. The units on the right hand side could be mass m, expressed in grams, and acceleration a, expressed in cm/sec^2. In that case the force has the units of $gm \times (cm/sec^2)$. This unit of force has a name; it is called a dyne. In working any problem in which the force is to be expressed in dynes, the unit of mass must be grams and the unit of acceleration must be cm/sec^2. Now suppose that in the same equation the units of mass are kilograms and the units of acceleration are $kilometers/sec^2$. What are the units of force?

Occasionally an equation contains a constant, sometimes referred to as a constant of proportionality. An example is the constant G in the law of gravitation. Since all of the other quantities in the law have physical dimensions, the constant must also have physical dimensions, and its value will depend on the set of units used to measure the other quantities in the equation. This is where a constant like G differs from a dimensionless constant like π, the ratio of the circumference to the diameter of a circle. If you are not sure if a constant has dimensions or not, try balancing the units in the equation. If they balance without any help from the constant, then the constant is dimensionless. For example, consider the formula for the area of a circle, $A = \pi r^2$. If r is measured in meters and area is measured in square meters, then π is dimensionless.

If you pay attention to balancing the units in an equation, it will be clear which ones should be used. Try the following problems for practice.

1. In the law of gravitation, there is a constant usually abbreviated as G. Determine the units that this constant has if the force is in units of dynes, the mass in grams and the distance in centimeters. (Note that $F = Gm_1m_2/r^2$, so $G = Fr^2/m_1m_2$.)

2. Suppose that you are faced with calculating the gravitational force in dynes between the Earth and the Moon. You are given the masses in grams and the Earth–Moon distance in miles. What must you do to the units before solving the equation?

3. If the escape velocity of a planet is given by $v_e = 2GM/R$, what are the units of v_e? Assume M is the planet's mass in grams, R is its radius in cm, and that you have found the units for G in question 1.

Exercise 2: The Meaning of Acceleration

Because its units at first seem strange, many students are confused by the meaning of acceleration. In this exercise we will try to understand the units of acceleration. We stated that the units of acceleration are a distance divided by a squared time: ft/s^2, or m/s^2. What is the meaning of this? The best way to think about acceleration is to read the units as "meters per second, *per* second", or "feet per second, *per* second."

It's easiest to understand the meaning by example: Suppose you get into your *Ferrari*, and start accelerating at 50 ft/s^2. Your acceleration means that you are increasing your speed by 50 feet per second, per second. Thus, after one second you are travelling at a speed of 50 feet per second. With a constant acceleration, your speed will continue increasing by 50 feet per second, each second. Thus, after two seconds you are up to 100 feet per second. After three seconds you are up to 150 feet per second. And so on.

The acceleration of gravity near the surface of the Earth is 9.8 m/s^2; to simplify our calculations, let's call it 10 m/s^2. That means (neglecting air resistance) objects in free fall will accelerate towards the surface of the Earth at a rate of 10 meters per second, *per* second. Try the following problems for practice (you should neglect air resistance):

1. Suppose you drop a penny from the top of a tall building. (Do not try this experiment! See answer at end of chapter.) You time its fall, and you see that it hits the ground after falling for 6 seconds. (This would happen if you drop a penny from about a 50 story building).
   a. How fast is the penny falling one second after you drop it?
   b. How fast is it falling two seconds after you drop it?
   c. How fast is it going after six seconds (just before it hits the ground)?
   d. Convert its speed just before hitting the ground from meters per second into km per hour.
   e. Convert the speed from km per hour into miles per hour. Do you think it would hurt if the penny hit someone on the head?
2. A player (let's call him Dwight) can throw a fast ball well over 90 miles per hour. Suppose, as a contribution to your astronomy course, Dwight agrees to come to our class one day and throw a baseball *straight up* at 90 miles per hour (we'll meet outside that day). If you do the conversion (but we're doing it for you this time!), you'll find that 90 miles per hour is the same as 40 meters per second.
   a. How fast, and in which direction (i.e., up or down), will the ball be going one second after Dwight throws it up?
   b. How fast, and in which direction, two seconds after it is thrown?
   c. When (that is, how many seconds after it was thrown) will the ball reach its highest point (momentarily coming to a stop before it starts falling back down)?
   d. How fast, and in which direction, five seconds after it is thrown?
   e. When will the ball reach the ground? (That is, how long after it was thrown up?)

Self Test

1. The Earth's gravitational force extends, at least in principle, as far as
   a. the Moon
   b. the Sun
   c. the edge of the solar system
   d. the edge of the universe.

2. Kepler's discovery that planetary orbits are really ellipses, rather than circles
   a. is entirely consistent with Newton's theory of gravitation
   b. eliminated the need for epicycles and deferents, making calculations of planetary positions easier and more accurate
   c. both of the above
   d. none of the above.

3. If we double the distance of an artificial satellite that has been in a synchronous orbit around the Earth, its new period will be
   a. 1 day
   b. 1.6 days
   c. 2 days
   d. 2.8 days.

4. An astronaut in a spaceship orbiting the Earth at a height of 500 km
   a. is too far from the Earth to feel the effect of gravity
   b. is falling towards the floor of his spacecraft
   c. weighs only one–fifth his ground weight
   d. will feel weightless.

5. The true diameter of a planet may be determined when two of the following are known (choose two):
   a. the mass of the planet
   b. the radius of the planet's orbit in kilometers
   c. the planet's rotation period
   d. the angular size of the planet's observed disk.

6. While the Moon is the primary cause of tides on the Earth, the Sun also exerts a tidal influence. At which lunar phase(s) would you expect tides to be the most pronounced?
   a. First quarter
   b. New moon
   c. Full moon
   d. Both new and full moons.

7. If you went to the Moon, which of your physical characteristics would be guaranteed to change?
   a. mass
   b. height
   c. weight
   d. density.

8. Which (there may be more than one) of the following are examples of acceleration?
    a. a car going 60 km/hr along a straight road
    b. a car turning a corner
    c. a wagon rolling down a hill
    d. the Earth orbiting the Sun.

9. If the law of gravitation stated that the force between two bodies was inversely proportional to the cube of the distance between them,
    a. the force would be weaker between the Sun and Earth than it is now
    b. the force would be stronger between the Sun and Earth than it is now
    c. the force would be the same at the distance of the Earth, but weaker for planets farther from the Sun
    d. there is no way to tell without an experiment.

10. The escape velocity of a planet depends on its
    a. mass and temperature
    b. density and temperature
    c. mass and radius
    d. none of the above.

Answers To Selected Review Questions from the Text

3. The average of these values is 27°14', to the accuracy of the measurements. If a single measurement is made, random errors in the measurement process may make the measurement inaccurate. If several measurements are made and averaged together then the effects of random errors can be minimized. This is one reason why Tycho Brahe's measurements were so much more accurate than those of earlier astronomers.

9. Because gravitation obeys an inverse square law, and Jupiter is about 5 times further from the Sun than the Earth, the gravitational attraction between the Sun and Jupiter would be about 5^2 or 25 times weaker than that of the Sun and Earth *if* the two planets have the same masses. However, since Jupiter has about 300 times the mass of the Earth, the product of the masses of the Sun and Jupiter is about 300 times the product of the masses of the Sun and the Earth. Thus, the overall gravitational attraction between the Sun and Jupiter is about 300/25 or 12 times stronger than that between the Sun and Earth. Similarly, the gravitational attraction between the Sun and Saturn is found to be approximately the same as that between the Sun and Earth.

10. There is gravity everywhere. The astronaut in an orbiting spacecraft feels weightless because she is in *freefall*.

11. Kepler did not realize that his third law depends on the *sum* of the masses of the Sun and planets because, in the case of all planets, this sum is virtually the same. This is because the masses of the planets are so small compared to that of the Sun that they are virtually unnoticeable.

Answer to Exercise 1

1. Units of G are dynes×cm²/gram.

2. You will need the distance converted to units of centimeters.

3. The units of escape velocity are cm/s.

Answer to Exercise 2

1. a. Since the penny is increasing its speed by 10 meters per second, *per* second: it is falling at 10 m/s after one second.
   b. After two seconds it is falling at 20 m/s.
   c. After six seconds, it is falling at 60 m/s.
   d. We convert 60 m/s into km/hr:

   $$(60 \text{ m/s}) \times (0.001 \text{ km/m}) \times (3600 \text{ s/hr}) = 216 \text{ km/hr}$$

   e. We convert from km/hr into miles/hr:

   $$(216 \text{ km/hr}) \times (0.6214 \text{ mile/km}) = 134 \text{ mile/hr}$$

   A penny hitting your head at 134 mile/hr? It probably would kill you.

2. a. The acceleration of gravity changes the speed of the ball by 10 meters per second, *per second*. Since Dwight threw the ball *up* at 40 m/s, it will have slowed to 30 m/s after one second. It still is going up.
   b. After two seconds, it is still going up, but only at 20 m/s.
   c. After three seconds, it would be going only 10 m/s; thus, it would stop going up after *four seconds*, when its speed is zero.
   d. After five seconds, it has started back down and is travelling at 10 m/s.
   e. This problem has *symmetry*: the time going up and down will be the same. Thus, since it took four seconds for the ball to reach the top of its arc, it will be a total of *eight seconds* for the round trip.

Answers To Self Test

1–d, 2–c, 3–d, 4–d, 5–b–d, 6–d, 7–c, 8–b–c–d, 9–a, 10–c

4
Messages from the Cosmos: Light and Telescopes

The light that we see with our eyes is only a small part of the entire spectrum of light. We refer to the light that we can see as *visible light*, and we refer to the entire spectrum of all kinds of light as the *electromagnetic spectrum*. Virtually all our knowledge of astronomy comes from the study of the light emitted by objects in the universe; thus, it is important for astronomers to observe light from all regions of the electromagnetic spectrum.

Experimentally, light has been found to exhibit properties both of waves and of particles. This apparent dual nature of light led to the formulation of *quantum mechanics* in the early part of this century. Today, we say that light comes in little packets called *photons*, and that each photon is characterized by a wavelength and a frequency. Visible light has wavelengths ranging from about 4000 Angstroms (blue light) to about 7000 Angstroms (red light). Light with wavelengths longer than those of red light is called *infrared*; still longer in wavelength we find *microwaves* and *radio*. (Note: Radio is a form of light, not of sound! We "hear" radio only after a signal carried by radio light has been converted into a sound by the electronics in our sound systems.) As we go to wavelengths shorter than those of blue light we find *ultraviolet*, *X–rays*, and *gamma–rays*. Light is sometimes referred to as *radiation*. Note, therefore, that not all radiation is dangerous; while X–ray radiation certainly can be harmful, visible radiation is not.

All objects emit radiation simply by virtue of their temperature. The radiation produced because of an object's temperature is called *thermal* radiation. Thermal radiation is *continuous*, because it does not contain emission or absorption lines. The two laws of thermal radiation are: 1) Wien's law, which states a relationship between the temperature of an object and the wavelength of maximum emission of its light — hotter objects emit most of their radiation at shorter wavelengths; 2) the Stefan–Boltzmann law states that the total energy of light emitted by an object, per unit area of surface, is proportional to the fourth power of its temperature.

When we spread light into its component wavelengths with a prism, we sometimes see continuous radiation (e.g., a rainbow of color) with narrow lines where the light is missing. These dark lines are called *absorption lines*, and are caused by the absorption of photons by atoms lying between us and the source of the continuous radiation. If we look at a hot, rarefied gas we may see only a few thin, bright lines of light (color) with no continuous light. These bright lines are called *emission lines* and are the result of atoms emitting at specific wavelengths as electrons jump between allowed energy levels. Kirchhoff's rules summarize the conditions under which absorption or emission lines are formed.

Atoms are made up of a *nucleus*, which contains protons and neutrons, and electrons which "orbit" the nucleus. We put the word "orbit" in quotes because, like photons, electrons exhibit both wave and particle properties; it turns out that only certain "orbits" are allowed to the electrons. Each allowed "orbit" is characterized by a particular energy level, and each chemical element has its own distinct set of allowed "orbits". Emission lines are formed when an electron drops from an "orbit" with a higher energy level to a lower one; the frequency of the emitted photon of light directly corresponds to the difference in the electron energy levels before and after the transition. Absorption lines are formed when an atom absorbs a photon causing the electron to jump from a low energy level to a higher one. Because the energy levels are unique for each chemical element, the pattern of emission or absorption lines allows us to determine the composition of distant objects such as planets or stars. Since the pattern changes as the atoms are excited and ionized, we can also study the density and temperature of astronomical objects.

Another important source of information in astronomical spectra is the Doppler shift. When an object is moving towards or away from us, the entire spectrum will be shifted either to the blue or the red by an amount proportional to the speed of the star along our line of sight. The shift of the continuous spectrum is very difficult to measure, but if there are absorption or emission lines in the spectrum, then the velocity can be directly measured by comparing the wavelengths of line patterns in the spectrum of the moving object to known wavelengths for the lines at rest.

A telescope is essentially a light bucket, a device that can collect far more photons than the human eye. Instruments can be attached to telescopes to record photons of light in a variety of ways, from simple photographs to electronic detectors. The most common telescope design for ultraviolet, optical, infrared, and radio light is that of the reflecting telescope. Incoming light is reflected from a concave mirror, either spherical or parabolic, to a point in front of the mirror. In some cases this is the focus of the mirror and the recording instrument is placed there. In other designs the light is reflected off another, secondary mirror either back through a hole in the primary mirror or off to the side of the light path. Another somewhat less common design for professional telescopes is the refractor, which employs a lens as the light–collecting element.

Light can be measured (e.g., photographed or counted by an electronic detector) either directly as it is collected by the telescope or after passing through a filter which allows only particular portions of the spectrum to pass. Devices which measure the intensity of filtered or unfiltered light are called *photometers*. Alternatively, a spectrograph may be hooked up to a telescope in order to spread out the light spectrum so that we may see its component wavelengths, thus enabling us to identify such features as emission or absorption lines. In all cases the image or spectrum is stored on film or electronically for later study.

Since most exposures with a telescope require the object be observed continuously for several minutes to several hours, it is necessary for the telescope to move during the observation in order to keep the object in the field–of–view. This telescope motion is called *tracking* of an object; tracking compensates for the apparent motion of objects across our sky due to the Earth's rotation.

Sites for Earth–based telescopes are chosen on the basis of their clear, steady air and the absence of light pollution. Increasingly, telescopes are being put in orbit where they are above all of the Earth's atmosphere. Such space observatories offer several advantages: for example, they can observe in any wavelength region and are never subject to weather conditions. Space observatories are expensive, however, and the largest telescopes still are built on the ground.

Words and Phrases

atom
Angstrom
absorption line
blue–shifted spectrum
cassegrain focus
continuous radiation
coude focus
diffraction
Doppler shift
electromagnetic spectrum
electron
emission line
energy levels
excitation
eyepiece
focal length
focus
frequency (of radiation)
gamma rays
hertz
infrared radiation
ion
ionization
interferometry

Kirchhoff's rules
microwaves
nucleus
photometer
photon
polarized light
prime focus
quantum mechanics
radio
red–shifted spectrum
reflecting telescope
refracting telescope
resolution
spectrograph
spectroscopy
spectrum
Stefan–Boltzmann law
thermal radiation
ultraviolet
visible light
wavelength
Wien's law
X–rays

Exercise

In understanding lenses and telescopes, it is helpful to think about one "telescope" that most people have used: the camera. A camera generally is designed along the lines of a refracting telescope, so that light passes through a lens and is focused ultimately on film to record the picture. The following exercises can be done with most cameras, in case you want to verify your answers.

Imagine that you have a camera in which the lens can be left open to take long exposures. On a clear night you set it up pointing at the north star and expose the film for three hours. What will your developed picture look like? Now suppose you point the camera at the celestial equator and take another three-hour exposure. How will this picture differ from the last one? Finally, suppose you attach your camera to the tube of a large telescope with a clock drive. If the camera is again pointing toward the celestial equator and you take a three-hour exposure, what will the picture look like?

Suppose your camera lens has the misfortune to be splattered with a drop of black paint which covers about one quarter of the lens. Describe how the same scene photographed before and after this accident would look (if you want to try this experiment, use a cardboard wedge to cut off part of the image without touching the lens!)

Self Test

1. Which of the following is not a type of electromagnetic radiation?
   a. visible light
   b. radio waves
   c. X-rays
   d. sound waves
   e. infrared energy.

2. To determine what kinds of atoms the outer regions of a star are made of, astronomers would measure the star's
   a. continuous spectrum
   b. distance
   c. spectral absorption lines
   d. luminosity
   e. radius.

3. Our Sun emits its most intense radiation in what region of the electromagnetic spectrum?
   a. radio
   b. infrared
   c. visible
   d. ultraviolet
   e. X-ray.

4. The Doppler effect
   a. allows us to measure an object's velocity across our line–of–sight
   b. is a shift of an object's spectrum which depends on its velocity in the line–of–sight
   c. is a shift in an object's apparent position with respect to background stars
   d. is a shift in an object's spectrum which depends on its temperature.

5. The brightness of an object
   a. is inversely proportional to the distance between object and observer
   b. is inversely proportional to the square of the distance between object and observer
   c. is directly proportional to the square of the distance between object and observer
   d. none of the above.

6. Why can radio telescopes work during the daytime?
   a. radio waves are invisible
   b. the Sun emits very few radio waves in comparison to visible light
   c. radio astronomers don't use photographic techniques
   d. radio waves are not scattered in our atmosphere like visible light.

7. The main function of a telescope is generally to
   a. magnify star images
   b. separate light into its spectrum
   c. help us see radiation that cannot get through clouds
   d. collect more radiation from faint objects.

8. When you look at a star through a large telescope, you see
   a. uniform disk
   b. a point of light
   c. a disk with surface features
   d. a spectrum.

9. Which form of light travels through space at the fastest speed?
   a. visible
   b. infrared
   c. X–ray
   d. None of the above; all light travels at the same speed.

10. Why do we place telescopes on satellites in Earth orbit to measure celestial X–rays?
    a. to escape interference from artificial sources of X–rays
    b. because celestial X–rays cannot reach the ground
    c. celestial X–ray observations are too dangerous from the ground.

Answers to Selected Review Questions from the Text

4. The color of a star indicates its wavelength of maximum emission. From Wien's law, we know that this is related to the star's surface temperature. Colors with shorter wavelengths indicate hotter stars. For example, red stars (long wavelength) are cool, while blue stars are hot.

5. Everything emits light by virtue of having a temperature. People have a temperature of about 310 Kelvin (equivalent to 98.6°F or 37°C), and thus our wavelength of maximum emission is in the infrared. An infrared camera can therefore see the light emitted by people in the dark.

6. To calculate the wavelength of maximum emission for thermal radiation we use Wien's law:

$$\lambda_{max} = \frac{.29}{T} \text{ cm} = \frac{2.9 \times 10^7}{T} \text{ Å}$$

where λ_{max} is the wavelength at which more photons are emitted than any other wavelength (the "dominant" wavelength) and T is the temperature in Kelvin. For a star at 25,000 K, we find:

$$\lambda_{max} = \frac{2.9 \times 10^7}{25000} \text{ Å} = 1160 \text{ Å (ultraviolet)}.$$

Using the equation similarly for the other objects, we find wavelengths of 11,600 Å for a star at 2,500 K; 12.5 Å for the Sun's corona at 2 million K; and 80,645 Å for a human body at 310 K.

7. The intensity of light goes as the inverse square of the distance. Since Pluto is about 25 times further from the Sun than Mars, the intensity of Sunlight at Pluto is reduced by a factor of 25^2, or 625, from the intensity at Mars.

34

14. Although *Space Telescope* is not as large as some ground–based telescopes, it will still offer advantages even in visible light observations. It will not be affected by atmospheric seeing, since it will be above the atmosphere. Also, it will not be subject to weather–related interruptions.

Answers to Self Test

1–d, 2–e, 3–c, 4–b, 5–b, 6–d, 7–d, 8–b, 9–d, 10–b

5
The Earth and Its Companion

Because all of the planets were formed together from the same cloud of gas (as was the Sun; we refer to the cloud as the "solar nebula"), we can learn much about the planets and the solar system through the study of comparative planetology. We have come to understand much about the Earth as a planetary system through our studies of other planets and our understanding of the Earth, in turn, helps us to understand the rest of the solar system. In this chapter we study the Earth in some detail so that we will be able to apply our understanding throughout our study of the planets.

Earth is the third planet from the Sun, and one of the four *terrestrial* planets; the others are Mercury, Venus, and Mars. In contrast to the Giant planets (Jupiter, Saturn, Uranus, and Neptune) which are made primarily of light, gaseous elements such as hydrogen and helium, the terrestrial planets are made of heavier elements and have solid surfaces. The Earth is the only planet on which liquid water exists, and our atmosphere with 77% nitrogen and 21% oxygen (plus small amounts of other gases) is unique.

The atmosphere of the Earth forms a thin layer around the surface and is divided vertically into four zones according to temperature profile. The lowest layer, where most weather occurs, is called the *troposphere*; other layers are the stratosphere, mesosphere, and thermosphere. The global circulation of the atmosphere is energized by radiation from the Sun, and the patterns of circulation are largely determined by the Earth's rotation.

The Earth's interior can be explored through the study of seismic waves; we have identified four main regions: the solid inner core, the liquid outer core, the mantle, and the crust. The core is primarily made of heavier elements, like iron and nickel, than the outer layers of the Earth. This suggests that the Earth has undergone a process called differentiation, in which heavy elements sink towards the center. The fact that differentiation occurred indicates that the interior of the Earth once was fully molten. Today, only the outer core remains molten; currents generated in the liquid outer core are thought to be responsible for the magnetic *dynamo* which generates the Earth's magnetic field. High–energy particles coming from space, known as cosmic rays, become trapped in the magnetic field creating *radiation belts*; because these particles can cause genetic damage to living organisms, the belts are important as a shield. Particles entering the magnetic field also are responsible for such activity as aurorae.

The mantle of the Earth is very rigid, but in fact is thought to be a very slow–moving fluid. The fluid properties of the mantle allow for convection, in which warm material rises while cool material sinks. These convection currents in the mantle are responsible for continental drift. According to the theory of plate tectonics, the crust is made up of large moving plates that continually are in motion as though floating on the mantle; because the mantle flows slowly the rate of plate movement normally is measured in centimeters per year. The interactions at plate boundaries give rise to most of the surface features of the Earth.

The surface of the Earth is constantly recycled through processes of plate tectonics over a long time scale, and by erosion processes — water, wind, glaciation and biologic activity — over shorter time scales. The Earth's present atmosphere bears little resemblance to its early atmosphere. The most volatile gases, such as hydrogen and helium, escaped from the Earth very early in its history. Outgassing from the Earth's interior (volcanos, etc.) is the source of most of the material in our atmosphere, but the chemical composition is largely the result of interactions with biological organisms.

The age of the Earth is about 4.5 billion years. This age is established by the radioactive dating of rocks, a technique which allows us to determine the time since a rock solidified. Unfortunately, the oldest rocks found on Earth are not a good indicator of the age of the Earth because geological activity has melted and reformed virtually all surface material. Somewhat older rocks were returned from the Moon by the Apollo astronauts, but these still do not date to the origin of the Earth. In fact, the oldest known rocks are some of the meteorites, material which has fallen to the Earth from space. It is by dating meteorites that we learn the age of our solar system to be about 4.5 billion years; since we assume the entire solar system formed together, that also is the age of the Earth.

The Moon is the only body in the solar system, besides the Earth, upon which humans have walked. Twelve astronauts (two each from Apollo 11, 12, 14, 15, 16, and 17) visited the Moon between July 1969 and December 1972. Much of our understanding of the Moon comes from study of Moon rocks brought to the Earth by the Apollo missions. The most obvious features of the lunar surface are craters, mountains, and the relatively smooth maria.

The Earth is believed to have formed from the coalescense of many smaller particles known as planetesimals, but the origin of the Moon has long been a mystery. A recent hypothesis suggests that the Moon formed when a large planetesimal collided with the forming Earth, and was partially vaporized. This vaporized material formed a hot disk around the Earth and subsequently coalesced to form the Moon.

Words and Phrases

aerosols	maria
albedo	mesosphere
aurora borealis	metamorphic rock
compressional waves	ozone
continental drift	planetesimals
convection	plate tectonics
core	radioactive dating
craters (impact)	regolith
cyclone	rilles
differentiation	sedimentary rock
fission theory	seismic waves
igneous rock	stratosphere
ionosphere	thermosphere
magnetic dynamo	transverse waves
magnetosphere	troposphere
mantle	volatile gases

Self Test

1. Which of the following is a major constituent of the Earth's present atmosphere that was not part of the atmosphere of a few billion years ago?
    a. water vapor
    b. hydrogen
    c. oxygen
    d. carbon dioxide.

2. The earliest evidence for life on Earth dates back
    a. 3 million years
    b. 100 million years
    c. 3 billion years
    d. 1 billion years.

3. X–rays from the Sun's corona:
    a. are absorbed in the Earth's thermosphere
    b. cause meteor showers
    c. break apart ozone in the stratosphere
    d. generally reach the surface and fry us.

4. The Earth's atmosphere is composed mostly of
    a. oxygen
    b. carbon dioxide
    c. nitrogen
    d. hydrogen.

5. Which of the following types of rocks require erosion for their formation?
    a. igneous
    b. sedimentary
    c. basaltic
    d. none of the above.

6. The lunar maria
    a. are vast plains flooded with lava
    b. are high mountains scarred by impact craters
    c. are where the oldest Moon rocks were found
    d. cover the Moon uniformly.

7. The oldest Moon rocks are older than the most ancient rocks found on the Earth because
    a. the Moon was formed long before the Earth
    b. the radioactive elements used for dating are rare on Earth
    c. the oldest Earth rocks were destroyed (recycled) by erosion and continental drift
    d. the Moon formed from older material than the Earth did.

8. Evidence that the Moon does not have a molten core includes
    a. the lack of a lunar magnetic field
    b. seismic waves caused by quakes monitored during the Apollo landings
    c. both a and b.

9. Evidence that lunar craters are impact craters, rather than having been formed by volcanic eruptions includes
    a. the shape of the craters
    b. central peaks in many craters
    c. rays emanating from craters
    d. all of the above.

10. Features found only on the near side of the Moon are
    a. craters
    b. maria
    c. mountains
    d. rilles.

Answers to Selected Review Questions from the Text

4. The fact that the overall density of the Earth is considerably higher than the typical density of surface rocks tells us that the Earth must have a dense core; probably several times more dense than ordinary surface rock. This core is thought to have arisen through the process of differentiation in which heavy elements gradually sank toward the center of the Earth.

6. The age of the surface of the Earth refers to the age of the surface materials; this is the length of time the rocks have been in essentially their present form. These ages can be quite a bit younger than the age of the Earth itself. Typical surface rocks on the Earth may have ages of order tens or hundreds of millions of years (the time since these rocks were last melted and then solidified) whereas the Earth itself is about 4.5 billion years old.

8. The half–life is the time required for half of the material to decay away. Therefore, if half the element is gone, the rock sample must be one half–life old; for the rock described in this problem with a half–life of 5 million years its age is thus 5 million years. After two half–lives, only 1/4 of the original material remains; in this case, the rock is 10 million years old. After three half–lives 1/8 remains, after four half–lives 1/16 remains, etc. Thus, if 1/64 of the original amount remains the rock is six half–lives old, or 30 million years in this problem.

11. If the impacts have occurred uniformly over time, and if the maria have one–fourth as many craters, then the maria must be only one–fourth as old as the rest of the Moon's surface. However, dating of Moon rock samples indicates that the maria are only about half the age of the surface. This implies that the rate of impacts on the surface of the Moon has slowed as time has passed; most of the cratering took place early in the history of the solar system.

13. Although the Earth has presumably been hit and cratered as often as the Moon, erosion has obliterated all but a very few craters. In addition, small particles which form micrometeorite craters in the lunar surface tend to burn up as meteors in the Earth's atmosphere.

Answers to Self–Test
1–c, 2–c, 3–a, 4–c, 5–b, 6–a, 7–c, 8–c, 9–d, 10–b

6
The Inferior Planets: Venus and Mercury

Mercury and Venus are known as inferior planets because their orbits lie closer to the Sun than our own. Thus, they never venture far from the Sun in our sky. Venus can often be seen near the time of Sunset or Sunrise; Mercury, because it is closer to the Sun, is very difficult to see.

Venus is very similar to the Earth in physical characteristics such as size and density. Although it is closer to the Sun than the Earth, calculations prove that its surface temperature would be only slightly higher than those on Earth *if* it had a similar atmosphere. Thus, it was long assumed that surface conditions on Venus would be similar to those on Earth. Because a thick cloud cover prevents visual observation of the surface, it was not until Venus was studied by radio observations that we learned how drastically different its surface conditions are from our own. Indeed, Venus exhibits such extremes of surface temperature and pressure that even specially designed instrument packages (Soviet *Venera* landers) have survived there no more than an hour. The cause of the high surface temperature is the *greenhouse effect* created by the presence of vast amounts of carbon dioxide in its atmosphere.

Radar observations of Venus revealed that it rotates very slowly, and in a *retrograde* sense. This unusual rotation is believed to be the result of a major collision during the process of its formation from planetesimals. The slow rotation also contributes to the atmospheric structure and circulation. The atmosphere is relatively stable, with circulation centered about the *subsolar point*. Venus has three layers of clouds; the clouds are made up of sulfuric acid droplets. Atmospheric chemistry also is affected by cosmic rays from the Sun, since Venus has no magnetic field to shield it from charged particles.

The surface topography of Venus has been mapped by orbiting space probes (*Pioneer Venus*) with radar. There are three general types of terrain: rolling plains, highlands, and lowlands. There is some evidence for tectonic activity, and even some suggestion that there may be currently active volcanos. Thus, it is believed that the interior of Venus resembles that of the Earth. There is no evidence for large scale plate tectonics, however, suggesting that the crust of Venus is not broken up like that of the Earth.

Venus and Earth are believed to have been very similar in their early histories; both planets accreted from planetesimals of similar composition. Even the early atmosphere of Venus, produced by outgassing, should have been very similar to that of the Earth. The vast differences in present condition can be traced to subsequent evolution and the slight difference in the amount of solar energy received by the two planets. Because Venus is closer to the Sun, water was able to exist only in the vapor state rather than condensing to liquid as on Earth. Like carbon dioxide, water vapor contributes to the greenhouse effect and the temperature on Venus rose. On both planets, there are similar amounts of carbon dioxide; on Earth, carbon dioxide dissolved in liquid water and was deposited in rocks; on Venus, the carbon dioxide remained in the atmosphere further contributing to the greenhouse effect. The water vapor on Venus eventually was lost: molecules were split into hydrogen which escaped to space and oxygen which became incorporated into rock.

Mercury has been carefully studied by only one space probe — *Mariner 10*. It is the second smallest planet in the solar system (Pluto is smallest); indeed, it is not much larger than the Moon. Because it has no appreciable atmosphere, the surface temperature is quite hot on the side facing the Sun, and quite cold on the night side. Mercury's rotation is coupled to its revolution so that it spins exactly three times for every two orbits around the Sun. This coupling is the result of tidal forces which act to ensure that Mercury's heavy side — its mass is distributed unevenly — is always either directly towards or away from the Sun at perihelion.

Because Mercury has a magnetic field despite its slow rotation, it is thought to have a large, dense core. This indicates that Mercury not only is differentiated but has a larger fraction of heavy elements than any other planet. The surface of Mercury appears similar to that of our Moon, although the craters are flatter and not so closely spaced. There are prominent cliffs, called scarps, which may have resulted from a shrinking of Mercury after the crust had cooled and hardened. One gigantic crater — Caloris Planitia — dominates much of the landscape and may be the result of the impact which caused Mercury to be heavier on one side.

Words and Phrases

greenhouse effect

oxidation

retrograde rotation

scarps

spin–orbit coupling

subsolar point

synchronous rotation

terrestrial planets

Exercise: Planetary Probes

The time that it takes a planetary probe to reach its destination is determined primarily by Kepler's third law, rather than its rocket engines. It would take a prohibitively large amount of fuel to go directly to a planet, so the probe is placed in an orbit which will intersect the planet's orbit at the correct time and place. Once the probe has escaped the Earth, it is re–directed into an orbit about the Sun which is calculated to intersect the orbit of the target planet when the planet is there. This is the reason the timing of the launch is critical. The shape of the orbit must be elliptical, and in the simplest case the Earth represents one extreme and the planet is the other extreme of the orbit. For example, to send a probe to Mercury, one would put the probe in an elliptical orbit around the Sun with aphelion at 1 A.U. (the Earth's distance) and perihelion at 0.4 A.U. (the distance of Mercury from the Sun). To find out how long it would take such a probe to reach Mercury we use Kepler's third law:

$$(m + M)P^2 = a^3,$$

where m represents the mass of the probe, M the mass of the Sun, P the orbital period of the probe, and a the semimajor axis of its orbit. In this formula, mass is measured in solar masses, so the Sun has M = 1 and the rocket, m, can be ignored. The semi–major axis a is measured in A.U., and is equal to one half the longest axis of the elliptical orbit. What is this axis? Now solve for the period, P. Remember that the rocket need only complete one half an orbit to get from the Earth to Mercury.

Clearly a rocket may only be launched when its orbit will cause it to intersect with Mercury itself, not just the orbit of Mercury. Draw a diagram of the solar system showing the relative positions of Mercury and the Earth at the time of launch, and also at the time that the rocket arrives at Mercury.

44

Next apply this reasoning to Mars to find the time required to send a rocket there. Compare your answer with the launch dates and arrival dates given in Chapter 7 for Martian probes.

Self Test

1. The high surface temperature on Venus is directly the result of:
   a. its proximity to the Sun
   b. the fact that it always shows the same side to the Sun
   c. a greenhouse effect caused by its carbon dioxide atmosphere
   d. heat released by droplets of sulfuric acid.

2. The slow, retrograde rotation of Venus is probably the result of:
   a. a collision with a large planetesimal during the process of formation.
   b. a collision with an asteroid relatively recently in the history of the solar system
   c. being captured by the Sun, since Venus was not originally a planet
   d. nothing — its rotation is not unusual.

3. Clouds on Venus are composed of
   a. water vapor
   b. carbon dioxide
   c. sulfuric acid
   d. methane.

4. The atmosphere of Venus
   a. is relatively stable compared to the Earth
   b. circulates about the subsolar point
   c. is characterized by giant cyclones
   d. a and c
   e. a and b.

5. The solid surface of Venus has been mapped
   a. from the Earth in visible light
   b. from spacecraft with infrared sensors
   c. from the Earth and from space using radar
   d. from the Earth using ultraviolet light.

6. The largest surface feature on Mercury is:
   a. Mare Crisium
   b. Syrtis Major
   c. Olympus Mons
   d. Caloris Planitia.

7. Scarps on Mercury are probably due to:
   a. volcanic activity
   b. ejecta from craters
   c. crustal cooling and shrinking
   d. space debris.

8. The density of Mercury is
   a. greater than the Earth's because it contains a larger fraction of heavy elements
   b. less than that of the Earth because Mercury is smaller
   c. not yet known because no landings on Mercury have taken place
   d. 3.4 gm/cm^3.

9. The relation between Mercury's spin and orbital period is an example of spin-orbital coupling because
   a. Mercury keeps the same face to the Sun at all times
   b. the same spot on Mercury always faces the Sun at perihelion
   c. the planet rotates three times in every two trips around the Sun
   d. all wrong. Spin–orbit coupling involves a planet and its satellite, and Mercury has no satellites.

10. The surface of Mercury is similar to that of our Moon except that Mercury
   a. has no prominent maria
   b. has a weaker magnetic field than the Moon
   c. has a thick atmosphere
   d. has a smaller density than the Moon.

Answers to Selected Review Questions from the Text

2. Both Venus and the Earth have similar quantities of carbon dioxide, but on Venus this resides primarily in the atmosphere, whereas on Earth the carbon dioxide primarily is found in (carbonate) rocks.

3. In order to determine the mass of a planet, one must observe its interaction with another body. For most planets, this means observing a satellite. Kepler's third law then allows one to find the total mass of the planet and its satellite if the period and distance of the satellite are known. Since satellites are usually much smaller than their parent planet, one can then assume that this answer represents the mass of the planet. However, Venus has no satellites, so its mass could only be found from its interaction with asteroids and space probes.

4. We use Wien's law to calculate the wavelength of maximum emission from Venus:

$$\lambda_{max} = \frac{.29}{T} \text{ cm} = \frac{2.9 \times 10^7}{T} \text{ Å}$$

Plugging in 750 K to the formula we find:

$$\lambda_{max} = \frac{2.9 \times 10^7}{750} \text{ Å} = 38,700 \text{ Å} .$$

5. Venus is unlikely to have charged particle belts surrounding it because it does not have a significant magnetic field.

8. The intensity of light goes as the inverse square of the distance. Thus if the distance at aphelion is 1.5 times the distance at perihelion, then the intensity of Sunlight at perihelion is 1.5^2, or 2.25, times greater.

Answers to Self–Test:

1–c, 2–a, 3–c, 4–e, 5–c, 6–d, 7–c, 8–a, 9–c, 10–a

7
Mars and the Search for Life

Mars is the fourth, and most distant, of the small, rocky planets that cluster near the Sun. Beyond Mars, we cross the Asteroid Belt and enter the realm of the Giant Planets. Mars has two small, potato–shaped moons: Phobos and Deimos. These moons resemble asteroids both in appearance and in physical characteristics.

Telescopic observations of Mars, beginning in the 17th century, revealed similarities to Earth: a 24–hour day, seasonal variations, and polar caps. These similarities led some observers to speculate that life exists on Mars. By the early twentieth century, an entire folklore of canals, cities, and Martians had been invented. Visions of great Martian cities ended with the Mariner 4 flyby of Mars in 1965, but the search for life on Mars continued. The Viking landers failed to find conclusive evidence for life, but more refined experiments will need to be performed before the question of life on Mars is finally settled.

The Martian atmosphere is made up primarily of carbon dioxide, like that of Venus; it is far less dense than that of Venus, however, and only a weak greenhouse effect operates. Seasonal variations occur both because of the tilt of the axis, and because of Mars's varying distance from the Sun in its elliptical orbit. Massive dust storms occasionally blanket the planet during the southern summer. The polar caps are composed of both carbon dioxide and water ice. The Martian surface reveals evidence of past water flows, including dry river beds and flood plains. Yet, because of the low atmospheric pressure, liquid water cannot exist on Mars today. Water, now frozen in near–surface soil and rock and at the polar caps, may have flowed in the distant past, perhaps when climatic cycles made the atmosphere denser and the surface warmer.

The topography of Mars is noted for its massive volcanic systems, far larger than any on Earth. This suggests that there is little crustal movement so that volcanos tend not to move away from subsurface hot spots. Although volcanos testify to past tectonic activity it is generally believed, but not certain, that tectonic activity has ceased. The absence of a detected magnetic field, along with the apparent lack of present tectonic activity, suggests that Mars does not have a fluid core.

The reddish color of Mars is due to iron oxide in the surface soil and rocks. Both iron and silicon are present in high concentration, suggesting that Mars is not highly differentiated. The moderate average density of Mars also supports this conclusion. It is likely that Mars was molten for only a short time during its formation and then formed a thick crust.

Words and Phrases

chaotic terrain

Deimos

laminated terrain

Olympus Mons

organic molecules

permafrost

Phobos

Valles Marineris

Self Test

1. Which of the following surface features are found on Mars?
    a. large volcanos
    b. dry river beds
    c. craters
    d. sand dunes
    e. all of the above.

2. The southern polar cap on Mars melts almost entirely during its summer, but the northern cap shrinks only slightly during northern summer. This is because
    a. the northern cap is much larger
    b. Mars is closer to the Sun during the southern summer
    c. the southern cap is almost entirely CO_2 ice
    d. the winds are stronger in the north.

3. The Martian satellites Phobos and Deimos
   a. are large and spherical like the Moon
   b. are small and irregularly shaped
   c. show evidence of volcanic activity on their surface
   d. exert tremendous tidal forces on Mars.

4. Evidence that Mars has no crustal movement (plate tectonics) comes from
   a. seasonal changes in the polar caps
   b. the very large shield volcanos
   c. the presence of dry river beds
   d. the occasional dust storms that blanket the planet.

5. The *Viking* experiments searching for life on Mars found:
   a. conclusive evidence that life exists today on Mars
   b. conclusive evidence that life never has existed on Mars
   c. strong evidence that life existed on Mars in the past, but not in the present
   d. unexpected chemical activity on Mars, but no evidence of Earth–like life
      or organic molecules.

6. The process in which heavy elements sink to the center of a still–molten planet
   is called:
   a. separation
   b. reduction
   c. differentiation
   d. oxidation.

7. Olympus Mons is:
   a. a giant valley on Mars
   b. the name of a dry river bed on Mars.
   c. a volcano much larger than Mt. Everest
   d. one of the two Martian moons.

8. The Martian atmosphere is composed primarily of
   a. nitrogen
   b. water vapor
   c. carbon dioxide
   d. hydrogen.

9. The reddish color of Mars is due to
    a. dust particles in the air
    b. silicon dioxide
    c. iron oxide
    d. carbon monoxide.

10. The "canals" seen on Mars by early observers
    a. were an optical illusion; they do not exist at all
    b. turned out to be chains of craters
    c. faded during heavy dust storms in recent years
    d. were confirmed by the *Mariner 9* photographs.

Answers to Selected Review Questions from the Text

2. Both Venus and Mars have atmospheres dominated by carbon dioxide. The physical conditions, however, are very different. Venus has extraordinarily high atmospheric pressure which allows the carbon dioxide to produce a strong greenhouse effect. On Mars, the pressure is low and the greenhouse effect weak.

3. When observing Mars spectroscopically, the spectral lines formed in the atmosphere of Mars can be distinguished from those formed in the atmosphere of the Sun because of the big temperature difference between the Sun and Mars. Gases in the Sun are largely ionized, whereas those on Mars are in the form of molecules. There is more potential for confusion with absorption lines formed in the Earth's atmosphere, but here again we can rely on the difference in composition. Also, if we observe at times when Mars is moving radially with respect to the Earth, we may find that the Doppler effect separates lines of Mars and Earth. Finally, we can eliminate both sources of confusion by observing Mars from space, where there is no atmosphere to scatter light of the Sun and we are not looking through our own atmosphere.

10. The fact that Mars has little or no magnetic field means that it cannot shield its surface from charged particles coming from the Sun. The Earth is protected from these particles by its magnetic field and, if this protection did not exist, the particles could cause harmful or fatal mutations in living organisms. If humans some day live on Mars, some form of protection will be required.

Answers to Self Test

1–e, 2–b, 3–b, 4–b, 5–d, 6–c, 7–c, 8–d, 9–c, 10–a

8
Jupiter: Giant Among Giants

Jupiter, the fifth planet from the Sun, is more than twice as massive as the rest of the planets combined. More than a thousand Earths could fit within Jupiter. Like the Sun, Jupiter is composed mostly of hydrogen and helium, but its mass is too small to allow energy–generating nuclear fusion. Jupiter has at least 16 moons. The four largest, Io, Europa, Ganymede, and Callisto, were discovered by Galileo, and can be seen with binoculars or a small telescope. Ganymede is the largest moon in the solar system — larger, in fact, than two of the planets: Mercury and Pluto. Jupiter also has a thin ring, too dim to be seen from the Earth. A great deal of our knowledge of Jupiter comes from the *Pioneer* and *Voyager* probes.

Jupiter has no solid surface to be seen, and images reveal its atmosphere to have a layered appearance. Dark–colored belts, regions of descending gas, alternate with light–colored zones where gas is rising. The direction of flow east or west also alternates from each belt or zone to the adjacent one. The banded appearance is created by the rapid rotation of Jupiter. The atmosphere also displays a wide variety of spots and other storms, the most prominent being the Great Red Spot which is known to have persisted for at least 300 years.

The center of Jupiter probably contains a rocky core of heavy elements which sank through the process of differentiation, but most of the interior is believed to be in the form of liquid metallic hydrogen. Above this zone, but below the cloud layer at the surface, is a region of ordinary liquid hydrogen. Thus, while Jupiter is often referred to as a "gas" planet, it really is mostly in the form of a liquid. The liquid interior, combined with the rapid rotation, creates a magnetic dynamo that powers Jupiter's strong magnetic field. Within Jupiter's magnetosphere there are very intense radiation belts with high densities of charged particles. These belts are the source of synchrotron radiation from Jupiter. The material in the belts is supplied by volcanic activity on Io.

The Galilean satellites, comparable to our Moon in size, show increasing geologic activity from the outermost one inward. Callisto, the farthest out from Jupiter, has an ice surface covered with craters. Ganymede shows evidence for ancient tectonic activity, which has erased some of its craters. Europa resembles a marble, cracked internally, but with a smooth surface. Io shows no impact craters, and is the most volcanically active body in the solar system. The source of heating for Io's interior is tidal stress resulting from an orbital resonance with Europa combined with the strong tidal force of Jupiter.

Words and Phrases

belts and zones
Callisto
Europa
Ganymede

Great Red Spot
Io
Io torus
liquid metallic hydrogen
synchrotron radiation

Exercise: Measuring the Mass of Jupiter.

Even a small telescope will show the four moons of Jupiter that were discovered by Galileo. From a single observation it is difficult to tell them apart, but repeated observations over a period of weeks will untangle their orbits. The outermost, Callisto, is the easiest to identify, and it can be used to compute the mass of Jupiter.

The diagram shows Jupiter and its four moons as they would appear on 17 successive nights. Keeping in mind that Callisto has the largest orbit, identify this satellite on as many nights as you can. You can then estimate the period (in days).

We now want to use Kepler's third law, as modified by Newton, to find the mass of Jupiter. The law can be written

$$P^2 G(m_1 + m_2) = 4\pi^2 a^3 \; ,$$

where P = period, a = separation (semimajor axis of Callisto's orbit), and (m_1 + m_2) = the total mass of Jupiter plus Callisto. The term G is the gravitational constant, equal to 6.67×10^{-8} dyne cm^2 gm^{-2}. All the other units must be expressed in the same system, so the period must be in seconds and a in centimeters, and the mass will be in grams.

Before solving Kepler's third law, we need to find the separation a. If we measure the maximum angular separation between Jupiter and Callisto, then we can use the distance of Jupiter from the Earth to compute the true separation. The answer is about 1.9×10^{11} cm.

Now solve the equation for $(m_1 + m_2)$. Clearly the mass of Callisto is much less than that of Jupiter, so your answer is really the mass of Jupiter. Compare your value with the one in table 8.1 of the text.

Positions of the Galilean Satellites

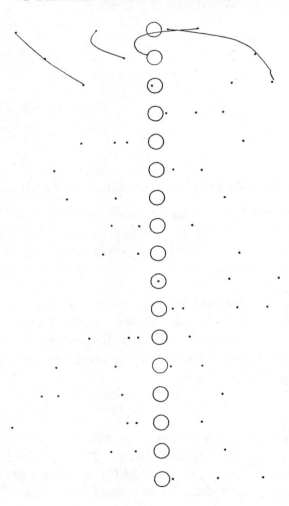

Self Test

1. Jupiter has a ring about it which
   a. was discovered through observations of radio synchrotron emission
   b. was discovered by the *Voyager* spacecraft
   c. lies outside the four Galilean satellites
   d. wrong: Saturn, not Jupiter, has a ring.

2. The primary source of material for Jupiter's intense radiation belts is:
   a. the Great Red spot
   b. particles from near Jupiter's poles
   c. Io
   d. the Sun.

3. The Galilean satellites are
   a. Phobos, Deimos, Io and Callisto
   b. Amalthea, Europa, Ganymede and Io
   c. Io, Europa, Ganymede and Callisto
   d. Ananke, Io, Europa and Ganymede.

4. The bands on Jupiter can be explained by a combination of
   a. cyclonic flow and convection
   b. differentiation and convection
   c. circulation and separation
   d. convection and rotation.

5. The mass of Jupiter is about
   a. 30 times that of the Earth
   b. 100 times that of the Earth
   c. 300 times that of the Earth
   d. 1 million times that of the Earth.

6. Jupiter rotates differentially, which means that
   a. it rotates faster at the equator than at the poles
   b. we see only light elements on the surface since heavy elements have sunk to the core
   c. the planet is flattened at the poles
   d. it has a magnetic field which extends very far from the planet.

7. In a sense, Jupiter is more like the Sun than the Earth, in that
    a. its composition is mainly hydrogen and helium
    b. it is so massive that nuclear reactions occur in the core
    c. the Great Red Spot, like Sunspots, is a region of intense magnetic field strength
    d. it is nearly the same size as the Sun.

8. The Great Red Spot of Jupiter is thought to be
    a. an extremely large cyclonic storm
    b. a volcano on the surface
    c. a simple cloud pattern which moves randomly
    d. a magnetic storm.

9. Jupiter's rapid rotation contributes to all of the following except:
    a. its oblate shape (equatorial diameter greater than polar diameter)
    b. its tidal effect on the Galilean satellites
    c. its strong magnetic field
    d. the circulation pattern of its atmosphere.

10. The satellite Io is interesting since it is the only body other than Earth on which we have seen:
    a. a mainly nitrogen atmosphere
    b. oceans with liquid water
    c. current volcanic activity
    d. primitive life.

Answers to Selected Review Questions from the Text:

2. The sum of the masses of the 8 planets other than Jupiter is 7.7×10^{29} grams. Jupiter's mass is 1.9×10^{30} grams, or about 3 times that of the other planets combined.

3. On both the Earth and Jupiter, the forces that govern atmospheric motions are basically the same: heating which causes convection and rotation of the planet. On Earth, the heating is caused by the absorption of solar radiation at the surface and in the ozone layer. On Jupiter, the primary heating source is its own interior. The rotation of Jupiter is much more rapid than that of the Earth, and this causes the flow patterns in its atmosphere to be stretched out into belts and zones instead of in the circular flow patterns seen on Earth.

11. Io has such great volcanic activity because its interior is kept heated by tidal stresses. Jupiter exerts a strong tidal force, but this alone cannot explain the volcanic activity. Rather, it is the interaction of the tidal force from Jupiter combined with a varying tidal force from the moon Europa which causes internal heating in Io.

12. The Io Torus is a belt of gas that follows Io in its orbit around Jupiter. The source of this gas is the volcanic eruptions on Io. The gas is ionized by collisions, and the charged particles then are subject to following the magnetic field of Jupiter. Because the magnetic and rotational axes of Jupiter are offset, the Io torus wobbles.

Answers to Self Test:

1–b, 2–c, 3–c, 4–d, 5–c, 6–a, 7–a, 8–a, 9–b, 10–c

9
Saturn and Its Attendants

Saturn, the sixth planet from the Sun, is the second largest planet in the solar system. Easily visible to the naked eye, Saturn is the most distant of the five planets ("wandering stars") that were known to the ancients. To anyone who has ever seen it through even a modest–sized telescope, Saturn imparts a feeling of wonder and delight. The magnificence of Saturn's ring system only increases when we examine close–up pictures returned by space probes. Although we now know that all of the giant planets have ring systems, the rings of Saturn are the brightest, probably the most complex, and certainly the best–observed.

Most of the physical characteristics of Saturn are quite similar to those of Jupiter: rotation, composition, atmospheric circulation, structure of magnetic field and radiation belts, etc. But there are some important differences. The slightly lower temperature on Saturn has allowed the formation of more complex molecules, and has permitted a wider cloud layer than that of Jupiter. As a result, the belts and zones on Saturn are more obscured and muted in appearance. The belts extend much closer to the poles than they do on Jupiter, probably because of seasonal variations in Sunlight on Saturn (Jupiter's rotation axis is scarcely tilted so Jupiter, unlike Saturn, does not have significant seasonal variations in solar energy input). Saturn has relatively less helium in its atmosphere than Jupiter, suggesting that helium has sunk. It is believed that interior differentiation still is occurring on Saturn; as material sinks it releases gravitational energy which provides a source of interior heating.

The current count for satellites of Saturn is 19, many of which were discovered by the *Voyager* missions. There is one giant moon — Titan, seven of intermediate size, and the rest are tiny. Several of these small moons have unusual orbital configurations. Some of the tiny inner satellites play a role in controlling the structure of the ring system, while the major satellites are interesting for other reasons. Titan has a dense atmosphere of nitrogen and methane. It is not known why Titan has such a thick atmosphere. The surface of Titan is obscured by atmospheric haze, so little is known about it; there is speculation that conditions on Titan may allow for liquid methane "lakes" and even methane ice or snow. All of the intermediate–sized satellites have low densities indicating a composition primarily of ices. Each has unique characteristics, and all pose fascinating questions.

Saturn's ring system consists of particles ranging in size from microscopic to a few meters across. Each ring particle orbits Saturn independently; in accord with Kepler's third law, those nearer the planet travel faster. The rings all lie within the Roche limit, meaning that no large satellite could have formed at their location because of tidal forces.

Some of the gaps in the ring system can be attributed to the gravitational forces of the satellites. *Voyager 2* revealed far more structure in the rings than is visible from Earth, making the ring system appear to be grooved like a phonograph record. The complex smaller structures are probably due to a variety of causes, ranging from small "shepherd" satellites to Saturn's magnetosphere. The magnetic field is almost certainly responsible for the spoke–like structures, seen in *Voyager* photographs, which are apparently caused by very small particles suspended out of the plane of the rings.

Words and Phrases:

Cassini division	shepherd satellites
metallic hydrogen	spokes (in the rings)
Roche limit	Titan

Exercise: Saturn's Rings

The nature of Saturn's rings was understood long before the *Voyager* fly–bys. Evidence that they consist of many small chunks of material, each in orbit about the planet, came from a variety of observations. The fact that stars could be seen through the ring system argued that the rings were not solid. Furthermore, a solid ring would be shattered by differential gravitational forces. It can be shown theoretically that a satellite with the same density as Saturn would be broken into small fragments by Saturn's tidal force, if the distance of the satellite from the center of the planet were less than 2.4 times the planet's radius. From Figure 9.1, measure the distance of the ring system from Saturn in units of Saturnian radii. How close to the planet is the outer edge of the system? How close is the inner edge?

Spectroscopic studies prove that the rings are composed of individual fragments in Keplerian orbits. Remember that the Doppler shift in a spectral line tells us the velocity of approach or recession. Consider the sketch of Saturn below and answer the following questions about the Doppler shift that you would expect to observe at the labeled points.

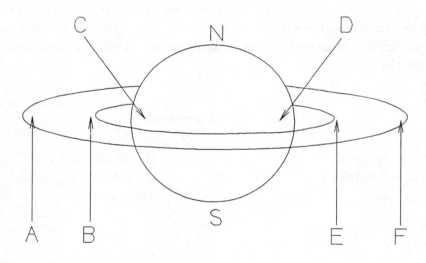

a. Suppose Saturn and its ring were not rotating at all. What Doppler shift would you measure at each point?

b. Suppose Saturn were rotating prograde (counterclockwise as seen from the north pole) and the rings were not rotating. What shifts would you see?

c. Suppose Saturn and its ring were rotating prograde, but the ring was a solid disk, like a phonograph record. What Doppler shift would you record?

d. Finally, suppose that the rings are indeed particles in Keplerian orbits. The inner ones will then be moving faster than the outer ones. What Doppler shifts will you see?

Self Test

1. The bands and zones on Saturn are not as pronounced as those on Jupiter because
   a. Saturn is farther away from the Earth
   b. Saturn has a thicker cloud layer overlying the bands
   c. Saturn is colder
   d. Differentiation of the elements is still going on.

2. Saturn's rings were first seen by
   a. early Greek observers
   b. Copernicus
   c. Galileo
   d. Newton.

3. Which of the following does not contribute to the ring structure of Saturn?
   a. orbital resonance with large satellites
   b. shepherd satellites
   c. spiral density waves caused by gravitational disturbances
   d. action of Saturn's magnetosphere on charged ring particles
   e. frequent collisions with large asteroids.

4. Saturn radiates more energy than it receives from the Sun. The primary explanation is that
   a. nuclear reactions are going on in the core
   b. Saturn still retains primordial heat from its formation
   c. differentiation is still going on in the interior
   d. none of the above. Saturn does not radiate more energy than it receives.

5. Which of the following statements about the ring system is not true?
   a. they orbit in the equatorial plane of Saturn
   b. they lie in a region where tidal forces would break apart any large satellite
   c. each individual ring is solid in structure, like a piece of string stretched around the planet
   d. the rings seen from Earth are in fact composed of many ringlets
   e. some of the rings are non–circular in shape.

6. Saturn experiences seasons while Jupiter does not because
   a. Saturn has a significant axial tilt and Jupiter does not
   b. Saturn's orbit is far less circular than Jupiter's
   c. Saturn is rotating faster than Jupiter
   d. wrong: Neither Jupiter nor Saturn experience seasons.

7. Saturn is not perfectly spherical, but bulges at its equator because
   a. the gravitational attraction of the rings distorts the upper atmosphere, which is what we see
   b. tidal forces from the large moon Titan cause the bulge
   c. the planet is rotating rapidly, causing the oblate shape
   d. seasonal variations distort its shape due to changes in the input of solar energy.

8. The origin of Saturn's rings is not known. All of the following origins are possible except:
   a. they are material deposited near Saturn by volcanic activity on Titan or other large moons
   b. they are the remains of a satellite which wandered too close to Saturn and was destroyed by tidal forces
   c. they are debris left over from the time that Saturn formed which was unable to coalesce into moons
   d. they are relatively young and are material from former moons that were destroyed by impacts.

9. Which of the following statements about Titan is not true?
   a. Its atmosphere is made mostly of nitrogen
   b. It is the largest moon in the solar system
   c. Our view of the surface is obscured by atmospheric haze and smog
   d. Its surface temperature is too low to allow for liquid water, but may allow for liquid methane.

10. Which Saturnian satellite has a high albedo on one side and a low one on the other?
    a. Enceladus
    b. Tethys
    c. Iapetus
    d. Mimas.

Answers to Selected Review Questions From the Text

3. Saturn's orbital period is almost 30 years. Since the rings of Saturn are inclined to its orbit (they lie in its equatorial plane), we will seen them edge–on to Saturn only when the ring plane lies directly in our line–of–sight. This will occur only twice in every Saturnian orbit, or about once every 15 years.

6. Methane on Titan is comparable to water on Earth in that the atmospheric conditions are near the so–called triple point, the point where a substance can exist in either liquid, solid, or gas phases. Thus, on Earth, water exists as liquid in oceans and lakes, solid in ice caps, glaciers, or snow, and gas in clouds. Similarly, methane might exist in all three phases on Titan.

9. The Cassini division in the rings of Saturn lies at a position where orbiting particles would have exactly half the orbital period of the satellite Mimas. Thus, this gap in the rings is an example of an orbital resonance in which Mimas controls the ring structure. This is similar to the orbital resonance between Europa and Io that leads to Io's volcanic activity.

Answers to Self–Test:

1–b, 2–c, 3–e, 4–c, 5–c, 6–a, 7–c, 8–a, 9–b, 10–c

10
The Outer Planets

The outer planets Uranus, Neptune, and Pluto all were discovered only in modern times. (Although Uranus is barely visible to the naked eye, it was not recognized as a planet by the ancients.) Following the discovery of Uranus, studies of its orbital properties led researchers to deduce the existence of Neptune; thus, Neptune was discovered through the mathematics of Newton's laws and its telescopic detection only served to confirm the discovery. Though the search for a ninth planet was similarly motivated by observed perturbations in the orbit of Neptune, its discovery was in fact accidental since Pluto is far too small to affect the orbit of Neptune.

Uranus and Neptune both belong to the class of giant planets along with Jupiter and Saturn. Pluto, on the other hand, is a misfit among the planets. Indeed, Pluto bears greater resemblance to the icy satellites of the outer planets, or to an asteroid, than to a planet itself. Much of our knowledge about Uranus and Neptune comes from the visits by the *Voyager 2* spacecraft in 1986 and 1989, respectively. There are no current plans for any space probes to Pluto.

Both Uranus and Neptune have ring systems, as do Jupiter and Saturn. Uranus and Neptune both have somewhat unusual satellite systems. Uranus, along with its satellites, rotates about an axis which lies nearly perpendicular to the ecliptic plane. Neptune's two known satellites, prior to the Voyager encounter, both are in unusual orbits. Triton orbits in a retrograde sense, while Nereid has a very eccentric orbit; six additional satellites discovered by Voyager are in normal, prograde orbits. Both planets probably experienced some type of collisional encounter to cause these unusual orbital patterns.

Little is known about Pluto because of its small size and great distance. Indeed, even its mass was unknown until the recent discovery of its moon Charon which enabled us to obtain a mass estimate through application of Kepler's third law. Pluto's orbit is the most eccentric of any planet, and actually brings it nearer the Sun than Neptune for about 20 years out of each 248 year period; we are in such a period at present, and Neptune is actually the farthest planet from the Sun during the years 1979 through 1999.

Words and Phrases

Ariel	Oberon
Charon	Pluto
Herschel	Titania
Miranda	Triton
Neptune	Umbriel
Nereid	Uranus

Exercise: A Scale Model of the Solar System

Distances within our solar system are so great that they can be difficult to comprehend. By creating a scale model of the solar system we can gain a better feel for the incredible distances between planets, the difficulty of planetary exploration, and the emptiness of space. In the table below, fill in the true diameters of the planets (and Sun) and their true distances from the Sun; you can get these values from the chapters in your text or from the appendices. Then compute the scaled diameters and distances, using a scale of *1 to 10 billion*; that is, you can find the scaled values simply by dividing the true values by 10 billion (10^{10}). (Note: this is the same scale used in an actual model of the solar system which is built on the University of Colorado, Boulder campus.) In order to make sure that you understand your results, you should give the scaled diameters in units of either millimeters or centimeters, and the scaled distances in units of meters.

	Diameter	Scaled Diameter	Distance	Scaled Distance
Sun				
Mercury				
Venus				
Earth				
Mars				
Jupiter				
Saturn				
Uranus				
Neptune				
Pluto				

Self Test

1. The planet with the most extreme tilt of its axis is:
   a. Mercury
   b. Earth
   c. Uranus
   d. Jupiter.

2. Pluto is a "misfit" planet in the solar system for all of the following reasons except:
   a. it is the smallest of all planets, yet is located in the part of the solar system dominated by giant planets
   b. its orbit is very eccentric, and significantly inclined to the ecliptic plane
   c. its surface temperature is very low, and it has a moon
   d. its composition is unlike that of any other planet, though it resembles some moons and asteroids.

3. Which of the following statements concerning the unusual rotation of Uranus is not true?
    a. it causes extreme seasonal variations in solar energy input
    b. the orbital plane of its rings and major satellites is inclined at the same angle as the equatorial plane of the planet
    c. it causes the planet to have a strong magnetic field
    d. it is likely the result of some collision that took place before the planet had fully formed.

4. The composition and interior structure of Uranus and Neptune most resemble:
    a. Earth
    b. Titan
    c. an asteroid
    d. Saturn.

5. The most bizarre Uranian moon, which shows some features of a very young surface and others of a very old surface, is
    a. Umbriel
    b. Miranda
    c. Titania
    d. Oberon
    e. Ariel.

6. The rings of Uranus are difficult to see because
    a. Uranus is so far from the Sun
    b. the rings have a low albedo
    c. the rings are thinner than those of Saturn
    d. all of the above.

7. Neptune's satellites Triton and Nereid are unusual because
    a. they both are extremely large
    b. they both have unusual orbits
    c. they both have substantial atmospheres
    d. their composition is unlike any other bodies in the solar system.

8. The atmosphere of Uranus
    a. shows detailed cloud structures like those of Jupiter and Saturn
    b. is virtually featureless
    c. shows light and dark regions, similar to those on Mars
    d. is composed primarily of nitrogen compounds.

9. Neptune was discovered
    a. by analyzing perturbations in the orbit of Uranus
    b. accidently, while searching for comets
    c. by the ancient Greeks
    d. by Clyde Tombaugh.

10. The density of Pluto indicates that it is made primarily of
    a. rocky material, like Mercury
    b. very heavy elements, like iron and nickel
    c. ices
    d. hydrogen and helium.

Answers to Selected Review Questions from the Text

2. If the motion of Uranus as seen from the Earth is 0.0005″/sec, and it takes 1 hour 40 min (= 6,000 seconds) to pass in front of a star, then the relation velocity × time = distance gives the angular diameter of Uranus:

$$(0.0005''/\text{sec}) \times (6,000\text{sec}) = 3''$$

3. Since the rotation axis lies nearly in the plane of the ecliptic, the north pole of Uranus receives Sunlight for about half of its orbit. Thus, a day at the north pole of Uranus is about 42 years long. When the pole is pointed 90 degrees from the Sun, the equatorial plane intersects the Sun. Thus, at this time, a day at the equator would be equal in length to the rotation period of Uranus, or a little over 17 hours (actually, half of this 17 hours would be in daylight and the other half in darkness).

10. Pluto and Charon are in mutual synchronous rotation unlike any other planet–satellite pair because Charon is so large relative to Pluto and the separation between the two is so small. Therefore, Charon exerts sufficient tidal stress on Pluto to have slowed Pluto's rotation over time until the two became synchronous. Though many planets have moons which are in synchronous rotation, no other planet has a moon with sufficient tidal force to have brought its planet into synchronous rotation as well.

Answers to Self–Test

1–c, 2–c, 3–c, 4–d, 5–b, 6–d, 7–b, 8–b, 9–a, 10–c

11
Space Debris

While the Sun and the planetary systems dominate the solar system, the remaining space is by no means empty. A variety of debris can be found scattered about including asteroids, comets, and interplanetary dust.

Asteroids, also known as minor planets, range in size from less than 100 km in diameter to about 1000 km for Ceres, the largest. About 2000 asteroids have been discovered, but the total number may be closer to 100,000. Nevertheless, the total mass of the asteroids adds to only a fraction of an Earth mass. Most of the asteroids cluster between the orbits of Mars and Jupiter. Certain distances from the Sun (within the asteroid belt) are remarkably devoid of asteroids; these regions are know as Kirkwood's gaps and correspond to orbital resonances with Jupiter. The asteroids are believed to be formed of material which never coalesced into a planet, probably because of the gravitational influence of Jupiter.

Comets are believed to originate in the Oort cloud, a cloud of debris which is thought to orbit the Sun at a very large distance. It is thought that comets represent primordial material from the solar system. Occasionally, a piece of this debris is disturbed and begins to fall inward towards the Sun, becoming a comet. Most comets are on very long period orbits, so we see them only once; only rarely, when they experience further gravitational encounters with the planets, do comets end up on short-period orbits that allow them to be seen over and over like comet Halley.

When far from the Sun, comets are frozen chunks of icy materials mixed with dust. When one approaches the Sun, however, heating allows volatile gases to escape, creating the coma and the tail. Often, there are two distinct tails known as the dust tail and the ion tail. A comet bright enough to be seen with the naked eye typically appears every few years, although bright city lights render many comets invisible to most people. A comet is usually visible for several weeks or months, its motion relative to the stars easily noticed from night to night.

Solid particles of debris that can, in principle, enter the Earth's atmosphere are called *meteoroids*. Most are very tiny, and burn up in our atmosphere as *meteors*. Sometimes the Earth passes through a swarm of tiny debris and we see a *meteor shower*; this swarm generally is the remains of a short—period comet which has disintegrated and its particles spread out along its orbit.

Occasionally, a meteoroid is large enough to survive the trip through our atmosphere and can be found on the ground as a *meteorite*. Meteorites generally can be grouped into three categories: stony, iron, and stony-iron. Most of the stony meteorites are of a type called *chondrites* because of the small, spherical chondrules they show. While all meteorites are made of material that is quite old, the class known as carbonaceous chondrites are thought to be material unprocessed since the time the solar system formed. The Earth occasionally is struck by larger objects which may leave impact craters. Impacts probably have been as common on Earth as on the Moon, but processes of erosion tend to erase the evidence of impacts on Earth over time.

Although the space between the planets is far less dense than any human-made "vacuum" on Earth, it is nonetheless filled with microscopic dust grains and particles of gas. Although the density of the dust is very low its presence is fairly easy to detect because it scatters Sunlight; two well—known effects of this scattering are the zodiacal light and the gegenschein. Gas from interstellar space appears to stream through our solar system at a speed of about 20 km/s. This interstellar wind is actually the result of our Sun's motion through the interstellar gas that occupies the space between stars.

Words and Phrases

amino acids - *a complex organic molecule of the type that forms proteins*

asteroids *small irregular bodies orbiting the sun, primarily in the asteroid belt*

carbonaceous chondrites - *a meteorite with chondrules and abundance of carbon*

Ceres *The first asteroid discoved, "(5th planet)"*

coma *The extended glowing region that surrounds the nucleus.*

comets *an interplanetary body of loose rock & ice*

chondrites *a stone meteorite with chondrules*

chondrules - *a spherical inclusion in many meteorites usually composed of silicates*

dust tail - *dust particles pushed away from sun by force of light they absorb.*

fluorescence - *emission of light at a wavelength*

fireball (or bolide) *an extremely bright meteor*

gegenschein - *the diffuse glowing spot, seen on the ecliptic opposite the sun's direction, reflecting sunlight off interplanetary dust.*

Halley's comet *a periodical comet*

interplanetary dust *- fine particles inbetween planets*

ion tail - *formed of ionized gases*

Kirkwood's gaps *Narrow gaps in asteroid belt formed by orbital resonance w/ Jupiter*

meteors - *a bright streak of light created when a meteoroid enter the atmosphere*

meteor shower - *a period where meteors are seen with high frequency.*

meteorites (stony, iron, stony-iron) *a fallen meteoroid*

meteoroid - *a small interplanetary body*

nucleus (of comet) *the central dense concentration*

Oort cloud *where comets originate*

sublimation *process which solid ice converted to gas.*

zodiacal light - *a diffuse band of light along the ecliptic near sunrise and sunset.*

Self Test

1. Which of the following statements contrasting meteors from comets is *not* true:
   a. Meteors flash through the sky in a few seconds, while a comet will be visible in the sky for many weeks or months
   b. Meteors are generally sand-grain size, while comets are several miles in diameter
   c. If you go outside on a dark, clear night you likely will see hundreds of comets, but probably see no meteors
   d. Meteors are phenomena that occur within the Earth's atmosphere, while most comets pass millions of miles away from the Earth
   e. Meteor showers are caused by tiny particles orbiting the Sun; these particles were lost from comets that passed near the Earth's orbit.

2. The existence of interplanetary dust is revealed by
   a. the zodiacal light
   b. the gegenschein
   c. ocean floor material
   d. all of the above.

71

3. The oldest meteorites, believed to be primordial material from the solar system, are known as
    a. stony-irons
    b. iron meteorites
    c. chondrules
    d. carbonaceous chondrites.

4. The best description of the composition of comets from the list below is
    a. gravel embedded in frozen gases
    b. carbonaceous compounds
    c. nickle–iron compounds
    d. hydrogen and helium.

5. Which of the following statements about asteroids is not true?
    a. they are thought to be material that never coalesced into a planet
    b. most of them are located between the orbits of Mars and Jupiter
    c. they are made of rocky materials, as opposed to gases like hydrogen and helium
    d. their combined mass is nearly as great as the mass of Jupiter.

6. Asteroids can be detected on long-exposure photographs because
    a. they show up as streaks due to their orbital motion
    b. they are much brighter than stars
    c. they can be seen as resolved disks, whereas stars appear as points of light
    d. they are of different colors than stars.

7. Meteor showers occur
    a. following a major impact of a large object on the Earth
    b. when the Earth passes through the orbit of a former comet
    c. randomly, at unpredictable times
    d. every 29 days.

8. Bode's law proves
    a. that the asteroids were once a planet
    b. that Neptune has moved since the solar system formed
    c. that Pluto had to exist
    d. nothing.

9. The origin of comets is believed to be
    a. collisions among asteroids in the asteroid belt
    b. material spewed out by volcanic eruptions on Io
    c. a reservoir of debris orbiting the Sun at a great distance, known as the Oort cloud
    d. material entering the realm of our solar system from interstellar space.

10. There are far fewer impact craters to be found on the Earth than on the Moon. This is because:
    a. the Earth formed after the Moon, and the period of heavy cratering in the solar system already was over
    b. large objects that might leave an impact crater on the Moon tend to burn up in the Earth's atmosphere
    c. by luck, the Earth has been hit by fewer large objects than the Moon
    d. erosion processes on the Earth tend to erase the evidence of impacts over time.

Answers to Selected Review Questions From the Text

1. From Kepler's third law we can find the period of the asteroid, taking M_S to be the mass of the Sun, M_A to be the mass of the asteroid (which is negligible in comparison), and a to be the semimajor axis of the asteroid's orbit. Then the period P is found from

$$P^2(M_S + M_A) = a^3, \text{ where}$$

$M_S = 1$ solar mass, $M_A = 0$, and $a = 2.8$ A.U. Solving for P yields

$$P = \sqrt{(2.8)^3/(M_S + M_A)} = \sqrt{21.95/1} = 4.7 \text{ years.}$$

2. The study of asteroids, comets, and meteorites is of particular interest to scientists concerned with the history of the solar system because these bodies are examples of matter formed early in the solar system's history.

8. Alpha Centauri is about the same mass as the Sun, and about 300,000 AU distant. If a comet is located 100,000 AU from the Sun, then it is 200,000 AU from Alpha Centauri; that is, it is twice as far from Alpha Centauri. Since the force of gravity follows an inverse square law, the gravitational force from the Sun is 2^2 or 4 times as great as that from Alpha Centauri.

10. Meteors are visible when they burn up in coming through the atmosphere. Mercury, with no atmosphere, would be a bad place to watch for them. However, one could expect to find many meteorites on the surface; indeed, the presence of impact craters on Mercury attests to the impact of material from space.

Answers to Self–Test

1–c, 2–d, 3–d, 4–a, 5–d, 6–a, 7–b, 8–d, 9–c, 10–d

12
Adding It Up: Formation
of the Solar System

Our solar system is organized according to a number of general patterns which provide clues to its origin. For example, all planets orbit in nearly the same plane and in the same directions (prograde, or counterclockwise as viewed from above the Earth's north pole). Nearly all planets rotate in the same sense, with the exceptions of Venus and Uranus (and Pluto, which is a misfit in any case). Nearly all of the major moons in the solar system also orbit in their planet's equatorial plane and in the same prograde direction. The compositions of planets, moons, and asteroids follow a pattern from the rocky inner worlds to the outer planets, built of light elements, and their icy moons. Along with other evidence, the discovery of these patterns has led to the general acceptance today of the idea that our solar system *evolved*, with the Sun and all its orbiting companions forming together from a cloud of material known as the *solar nebula*.

Vague outlines of an evolutionary theory to account for the origins of our solar system date back as far as the seventeenth century and the writings of Descartes. The development of evolutionary theories was stymied, however, because they were unable to account for the slow solar rotation rate: angular momentum considerations lead to the prediction that the Sun should be rotating rapidly in any evolutionary scenario. As a result, a competing class of theories arose known as catastrophic theories, in which the planets are thought to arise from a cataclysmic event that disrupts the Sun. As they were studied, it was found that the catastrophic theories suffered a variety of difficulties in explaining observed characteristics of the solar system; today, they can be definitively ruled out because of their inability to account for the presence of deuterium in the solar system.

Meanwhile, evolutionary theories came back into favor with the discovery of the solar wind in the 1960's. The slow solar rotation now can be explained by a process known as *magnetic braking* in which charged particles in interplanetary space are caught in the rotating magnetic field of the Sun. As the magnetic field tries to pull them along, drag is exerted on the Sun thus slowing its rotation. This mechanism was most effective in the early history of the solar system, when the density of interplanetary gas was higher than today, and over billions of years it has slowed the solar rotation to its present rate. Further evidence for the evolutionary theories comes from observations of dense interstellar clouds where newly–formed stars are found and from theoretical modelling of the process by which a cloud of gas can collapse to form a solar system.

The modern scenario for the formation of the solar system envisions the collapse of a cloud of interstellar gas and dust. The initial collapse must have been caused by some "trigger", possibly the shock wave from a nearby supernova. As the cloud collapsed, rotation caused it to flatten into a disk-shape with the embryonic Sun at the center; we refer to this disk of a material as the solar nebula. Temperatures were high near the center of the disk, becoming quite cold at larger distances. Solid particles began to form throughout the solar nebula, but their composition differed. In the inner solar system, only *refractory* materials, those with high melting points, are able to condense. In the outer solar system, however, the more common volatile gases are able to condense as well, forming ices. These chunks of solid material, known as planetesimals, coalesced to form the planets. Thus, the inner planets are low in volatile gases because they condensed from rocky planetesimals; the small sizes of the planets, along with tidal forces and high temperatures due to proximity to the Sun, prevented these bodies from retaining light gases as atmospheres. In the outer solar system, however, the giant planets were able to gravitationally trap the gases around them and thus reflect the composition of the original solar nebula. The formation of eddies around the giant planets led to their systems of satellites. Finally, the remaining material in the solar system was swept out by the T Tauri wind of the young Sun, or remained to crash into large bodies at some time in the future (making impact craters). The strong solar wind also contributed to slowing the rotation of the Sun by magnetic braking.

If our understanding of the formation of the solar system is correct, there should be many other stars with planets. At present there is no way to detect such planets. However, steps leading to the formation of stars have been observed in cloudy regions of the galaxy, and infrared observations of many nearby stars indicate that they are surrounded by rings of material which might eventually condense to form planets. New instruments might allow detection of large planets in the near future.

76

Words and Phrases

accretion *- gas surrounding a compact object*

giant planets *4 Gas giant planets, low density & lightweight gases*

magnetic braking *charged particles caught in Sun's magnetic field*

obliquity *- Equatory plane nearly orbital planes*

planetesimals *a small solar system body believed to have combine to form planets. Orbital motion in normal direction*

prograde motion

refractory elements *- exist in solid form it cooled*

solar nebula *- originated our Solar System*

solar wind *a stream of charged subatomic particles extending from Sun.*

terrestrial planets *- made of rock & metallic*

T Tauri wind *- a strong wind that swept gas & tiny dust particles*

volatile elements *require low Temperatures in order to condense tended to exist in gas form.*

Self Test

1. The age of the solar system is about
   a. 4.5 billion years
   b. 4.5 million years
   c. 1 billion years
   d. 100,000 years.

2. According to the accepted theory of the formation of the solar system
   a. the planets formed much earlier than the Sun
   b. the planets and Sun formed at about the same time
   c. the planets formed after the Sun had expelled much of the solar nebula
   d. the planets formed from cometary debris.

3. If magnetic braking had not operated over the Sun's lifetime,
   a. the planets would all orbit the Sun more quickly
   b. the planets would be farther from the Sun than they are at present
   c. the planets would rotate faster
   d. the Sun would rotate faster.

4. Which of the following is *not* accounted for by the theory of solar system evolution, but instead must be considered to be the result of a special event (like a collision)?
   a. the rings of Uranus
   b. the rapid rotation of the giant planets
   c. the retrograde rotation of Venus
   d. the icy composition of Saturn's moons.

5. Which of the following patterns in the solar system are accounted for by current theories of solar system evolution?
   a. the fact that planetary orbits are nearly circular
   b. the fact that planet satellite systems resemble the solar system in miniature
   c. the prograde orbits of all planets
   d. all of the above.

6. The T Tauri wind is
   a. just another name for the solar wind today
   b. the wind of the young Sun, which is stronger than today's solar wind
   c. a wind from the planet Jupiter
   d. a wind which swirls through the solar system like a tornado.

7. The planets account for what fraction of the total solar system mass?
   a. over 90 percent
   b. around 10 percent
   c. less than 1 percent
   d. not well known.

8. Planetesimals are
   a. small bodies in orbit between Mars and Jupiter
   b. material found in comet tails
   c. small bodies that were the first to condense from the solar nebula
   d. particles making up the rings of the giant planets.

9. Refractory elements are
   a. elements that split easily during nuclear fusion
   b. elements that condense at relatively high temperatures
   c. elements that burn
   d. elements that bend light.

10. If the evolutionary theory of planetary formation is correct, which of the following statements is likely to be true of all other planetary systems?
    a. they have 7 or more planets
    b. the planets all orbit in the same direction around their star
    c. the inner planets will have no moons
    d. the planetary spacing follows Bode's law.

Answers to Selected Review Questions from the Text

3. When a large object with little rotation, like an interstellar cloud, collapses to a small size, like the Sun, conservation of angular momentum dictates that the small object should rotate rapidly. However, the Sun in fact spins quite slowly — only one rotation per month. Thus, evolutionary models could not be taken seriously until a mechanism for slowing the solar rotation was discovered (magnetic braking). This is a similar situation to the slow acceptance of the heliocentric theory. Though the idea of a Sun-centered solar system had many advantages over the prevailing geocentric view, there was no direct evidence to support it; indeed, the absence of apparent stellar parallax (because it is too small to be seen with the naked eye) was taken as evidence against the notion of a revolving Earth. Similarly, the theory of continental drift was not accepted until a mechanism for it was discovered, despite ample evidence from the "puzzle-piece" arrangement of continents, etc.

5. It is thought that the entire solar system formed from a cloud with a uniform composition of mostly hydrogen and helium and only small amounts of heavier materials. This is the composition of the Sun and giant planets today. The terrestrial planets, however, are quite different in being made primarily of the rare heavier elements. This is because: 1) the inner planets formed in a warm region which prevented the condensations of light gasses; and 2) because of low mass, high temperature, and tidal forces from the Sun, light gasses near these planets or released from their interiors were able to escape.

6. We know from the discussion in this chapter that the embryonic Sun emitted primarily in the infrared. This implies that a search for sources of infrared radiation inside of interstellar clouds might be a profitable way of searching for stars in the process of formation.

9. The brightness of Jupiter at the distance of α Centauri can be found from the inverse square law:

$$\frac{\text{intensity at } \alpha \text{ Cen}}{\text{present intensity}} = \frac{1/(270,000 \text{ A.U.})^2}{1/(4 \text{ A.U.})^2} = 2.2 \times 10^{-10}$$

So Jupiter would appear only about 2 ten billionths as bright if it were orbiting the nearest star. This might be barely detectable with the largest telescopes, but the proximity of the bright star which it orbits is likely to drown out the image of Jupiter. (Nevertheless, direct detection of planets by the Space Telescope is not out of the question.)

79

Answers to Self–Test

1–a, 2–b, 3–d, 4–c, 5–d, 6–b, 7–c, 8–c, 9–b, 10–b

13
The Sun

Our Sun is a typical star, with nothing remarkable about it, except of course that it provides us with our habitable living conditions and is our principal source of energy. It is unique in another human–oriented way as well: it is the only star close enough for us to examine in great detail. By studying the Sun we learn much about the properties of stars; conversely, by studying stars, we can learn much about our own Sun.

The Sun is a massive sphere of gas, composed roughly of 73 percent hydrogen, 25 percent helium and 2 percent of all the other elements (as percentages of the total mass of the Sun). The interior of the Sun is in *hydrostatic equilibrium*, with the gravitational pressure inward being balanced at all points by the outward pressure of the Sun's hot gas. At the center of the Sun, densities and temperatures are so great that atomic nuclei, stripped of their electrons, collide with energies sufficient to induce nuclear fusion. In each fusion reaction, four hydrogen nuclei are converted to one helium nucleus; in the process some of the mass is converted into energy in accord with Einstein's famous relation $E = mc^2$. The fusion of hydrogen in the Sun proceeds by a series of steps known as the *proton-proton* cycle. Most of the energy is in the form of photons which are quickly absorbed, and then re-emitted, by other solar material; thus the energy reaches the surface only slowly, taking perhaps a million years from the time it is created to the time it is released into space.

The visible surface of the Sun is called the photosphere, and is the layer from which most of the visible photons are released to space; the Sun has no solid surface. The photosphere has a temperature of about 6000 K, and is granulated in appearance; the granules bubble with convection. Sunspots usually can be seen in the photosphere, and represent cooler material in a region of intense magnetic fields. While individual Sunspots last just a few weeks, the number and location of Sunspots varies with the solar activity cycle; minimum numbers of Sunspots occur about every 11 years. Because the magnetic polarity of Sunspots reverses with each minimum, the complete magnetic activity cycle averages about 22 years. Sunspots are the sites at which solar flares, sudden and dramatic releases of energy, occasionally occur. Major solar flares send large quantities of charged particles streaming towards the Earth where they may cause such phenomena as aurorae or disruptions in radio communication.

The temperature rises both above and below the photosphere (the density, however, increases inward but decreases outward). Just above the photosphere is a layer called the chromosphere, with temperatures up to about 10,000 K. Above that is the corona, where temperatures reach over 1 million Kelvin. The region in which temperatures suddenly rise between the chromosphere and the corona is known as the transition zone. Features and structures in these outer layers of the solar atmosphere vary considerably with the solar activity cycle.

The solar wind is a continuous flow of charged particles outward through the solar system. Because charged particles tend to follow magnetic field lines, the solar wind is thought to emanate primarily from the open magnetic fields of coronal holes. The solar wind feeds and shapes the Earth's radiation belts, and may also affect the Earth in other ways. The solar wind also is responsible for forcing comet tails to point away from the Sun. Because of the many interactions between the Sun and Earth, an understanding of stars is important not only for pure knowledge but in order to understand our own planet as well.

Words and Phrases

chromosphere	neutrino
corona	photosphere
coronal hole	solar prominence
deuterium	proton–proton chain
differential rotation	solar activity cycle
fission (nuclear)	solar constant
fusion (nuclear)	solar flares
granulation	solar wind
hydrostatic equilibrium	Sunspots
limb darkening	supergranulation
luminosity	transition zone
Maunder minimum	Zeeman effect

Exercise: Solar Energy

The solar constant, the amount of power received at the Earth from the Sun, is about 1400 watts per square meter. That is, *above* the Earth's atmosphere (so that we do not have to worry about clouds or darkness) there are 1400 watts of solar power passing through each square meter. The current power requirements of the United States are about 10^{13} watts. How many square meters of solar cells would be required in order to power the United States by solar energy assuming that the solar cells are 100% efficient?

Real solar cells are only about 10% efficient at best; we would therefore require at least ten times more collector area than you calculated above. Convert your answer above into units of square kilometers, and then compare the total amount of collector area we would require (with real solar cells) to the total surface area of the United States (about 2×10^7 km^2). Would it be practical for us to replace all of our current energy sources with solar power? Explain.

Self Test

1. Which of the following is (are) not seen in the solar photosphere?
   a. Sunspots
   b. granulation
   c. spicules
   d. limb darkening.

2. When a large solar flare is seen, we can expect radio communication difficulties a few days later. The reason for this delay is that
   a. radio waves are the last form of energy to be created in a flare
   b. charged particles travel more slowly then electromagnetic radiation
   c. particles can only escape through coronal holes
   d. it takes several days for the flare to heat the ionosphere.

3. In comparison with the photosphere, the corona has
   a. higher temperature and lower density
   b. higher temperature and higher density
   c. lower temperature and lower density
   d. lower temperature and higher density.

4. If nuclear reactions were to suddenly stop in the center of the Sun, we would feel the effects about
   a. 2 seconds later
   b. 8 minutes later
   c. 25 days later
   d. 1 million years later.

5. The Maunder minimum
   a. is a point in the "butterfly diagram" where the number of Sunspots is zero
   b. is the point in the photosphere where the temperature is at its lowest
   c. was a period in the 17th century when virtually no Sunspots were seen on the Sun
   d. refers to the low number of Sunspots every 11 years.

6. Which of the following provides evidence that convective motion dominates the heat flow in the outer layers of the Sun?
   a. the surface granulation
   b. the temperature gradient from center to surface
   c. limb darkening
   d. the high temperature of the corona.

7. The magnetic field of the Sun is important in
   a. Sunspots
   b. shaping the structure of prominences
   c. controlling the flow of the solar wind
   d. all of the above.

8. The visible part of the solar photospheric spectrum shows
   a. thousands of absorption lines, many due to iron atoms
   b. emission lines indicative of extremely high temperatures
   c. that only a minority of the chemical elements known on Earth are present in the Sun
   d. that the Sun is hotter in its outer layers than in its interior.

9. The Sun produces energy by
   a. the fission of uranium
   b. the fusion of carbon into oxygen
   c. the fusion of hydrogen into helium
   d. chemical reactions changing hydrogen and oxygen into water.

10. The total lifetime of the Sun (including both past and future) is about
     a. 1 million years
     b. 10 million years
     c. 1 billion years
     d. 10 billion years
     e. 1 trillion years.

Answers to Selected Review Questions from the Text

1. Hydrostatic equilibrium is the balance between gravity pulling inward and gas pressure pushing outward. Because the inward and outward forces both are uniform in all directions, the Sun is spherical. Rotation creates a centrifugal force which tends to cause bulging near the equator or, equivalently, flattening near the poles. Because the Sun is rotating so slowly, however, no significant distortion in its shape has been detected.

2. Nuclear reactions require very high densities so that collisions between nuclei are frequent, and very high temperatures so that the collisions occur with sufficient energy to overcome the mutual electric repulsion of the positively charged nuclei. These conditions occur only near the center of the Sun.

3. If the Sun's luminosity were ten times greater then it would be using its nuclear fuel ten times faster. If it had twice as much fuel available (twice the mass), then the net result would be a lifetime one-fifth as long as the actual ten billion year lifetime, or two billion years.

6. The photosphere is a relatively cool gas in front of hotter, underlying layers. Thus, in accord with Kirchhoff's laws, it absorbs specific wavelengths from the continuous radiation coming from below. In contrast, the chromosphere and corona are hot and rarefied so that they produce emission lines.

8. The intensity of thermal radiation, per unit area, is proportional to the fourth power of the temperature. Thus, if the temperature of a Sunspot were $1/2$ as hot as the surrounding photosphere, the intensity of its radiation per unit area would be $1/2^4$, or $1/16$, as great.

Answers to Self-Test

1–c, 2–b, 3–a, 4–d, 5–c, 6–a, 7–d, 8–a, 9–c, 10–d

14
Observations and Basic Properties of Stars

Although stars appear merely as points of light to the naked eye, modern instruments have allowed us to deduce a great deal about the nature of stars through study of that light. Indeed, we now can determine a wealth of information about stars including their distances, motions, sizes and masses, composition, rotation, etc.

Stellar positions can be accurately determined and, for nearby stars, we can measure their distances directly through the observation of parallax. (Note: detection of parallax is a direct proof that the Earth goes around the Sun, and not vice versa.) Accurate measurement of position also enables us to measure proper motions and, sometimes, to observe the motions of stars in binary systems.

Stellar brightness is measured using a modernized version of the magnitude system of Hipparchus. The magnitude scale runs backwards, so that large numbers represent dimmer objects; the brightest objects have negative magnitudes. Each five magnitudes on the scale represents a factor of 100 in relative brightness. Apparent magnitudes measure the brightness of objects as seen here on Earth. Absolute magnitudes compare the true luminosities of objects, since they are defined as the magnitudes objects would have *if* they were located at a distance of 10 parsecs from Earth. The true luminosities of stars are found to vary from about 1/10,000 that of the Sun to about one million times brighter than the Sun. Magnitudes are sometimes measured while placing filters over the detector in order to collect light of only particular wavelengths. The comparison of stellar magnitudes measured with two different filters is called a color index, and its value often can be related to the temperature of the star.

The first stellar spectra were recorded photographically long before their physical significance was understood, but early efforts to classify stellar spectra led to a scheme which was later found to be based on surface temperature. The letters O, B, A, F, G, K, and M define a sequence from hottest to coolest stars, representing a range in surface temperature from about 50,000 K to 2000 K. Some stars vary in brightness or spectra over time, and are known as variable stars.

Double, or binary, stars are important both because they are very common and because of the information they yield. Binary stars can be detected astrometrically from their orbital motion; spectroscopically from the periodic (Doppler) shifts of their spectral lines; or, if the Earth happens to be aligned with the orbital plane of a system, its binary nature may be revealed by periodic eclipses. It is through the application of Kepler's third law to observed binary systems that we are able to measure stellar masses. Stellar diameters also can be determined by observing the eclipses of eclipsing binaries.

The Hertzsprung–Russell diagram is a scatter plot of stars with spectral type or temperature plotted on the horizontal axis (but plotted backwards, with higher temperatures to the left), and luminosity or absolute magnitude on the vertical axis. When such a plot is made for many stars, it is found that the stars are grouped into several distinct regions. The great majority of stars fall into a diagonal strip running from the upper left (bright and hot) to the lower right (cool and dim) called the main sequence. The main sequence is assigned luminosity class V; luminosity class I represents supergiants and III represents giants while the other roman numerals represent classes in between. If we determine the location of a star on the H–R diagram by classifying its spectrum, then we can determine its distance by comparing its observed (apparent) magnitude to the absolute magnitude known for stars of its type. This method of distance determination is called *spectroscopic parallax*.

Several additional properties of stars can be deduced from the analysis of their spectral lines. These include the rotational velocity, derived from the degree to which the lines are broadened by the Doppler effect; the chemical composition, usually deduced by computing synthetic spectra for comparison with observed ones; and the magnetic field strength, inferred from the line–splitting phenomenon known as the Zeeman effect.

Words and Phrases

absolute magnitude	magnitudes
apparent magnitude	main sequence stars
astrometric binary	parallax
astrometry	parsec
bolometric magnitude	proper motion
color index	pulsating variable star
distance modulus	spectral type
eclipsing binary	spectroscopic binary
giant stars	spectroscopic parallax
H–R diagram	visual binary
luminosity class	white dwarf star

Exercise: The Many Motions of a Star

We have discussed several ways in which a star may be observed to move: precession, parallax, and proper motion are the most important ones. Let us now examine the size of these motions and see what factors they depend on.

The earliest stellar motion noted by mankind was precession, which has nothing to do with the stars themselves, but is a shift of the stars with respect to our reference frame, the Earth. The period of precession is 26,000 years, so the intersection of the celestial equator and ecliptic will slide completely around the equator in this time. Calculate the rate of precession at the equator, using a shift of 360° in 26,000 years. Will this rate be the same in all parts of the sky? To answer this question start by asking yourself what the rate of precession will be at the ecliptic pole. Next consider the precession halfway between the pole and the equator.

Since parallax is so small compared to precession it took almost 2000 years longer to detect it. In speaking of parallax, we describe the star shifting back and forth against the background of more distant stars. In fact this is only an accurate description of what we see when the star is exactly in the plane of the ecliptic. Draw a sketch of what we see if the nearby star is close to the ecliptic pole. As a hint, remember that as the Earth moves around the Sun while we watch a star near the ecliptic pole, our line of sight traces out a cone. Finally, describe the parallactic motion of a star at intermediate latitudes between the equator and pole.

While the size of the parallax depends on the distance of the star, the proper motion of the star depends on both the distance and the star's intrinsic motion. Of the three motions discussed here, only proper motion will eventually alter the appearance of the constellations. Suppose we consider the Big Dipper. The two stars in the bowl which point to the north pole (the pointer stars) are Merak and Dubhe. Merak has a proper motion of 0.088"/year, directed 71° east of north. Dubhe has a proper motion of 0.138"/year, directed 240° east of north. Calculate how long it will take Dubhe to move 5°, approximately the present separation of these two stars. Where will Merak be then?

Self Test

1. Differences among stellar spectra are primarily due to differences in
   a. luminosity
   b. radius
   c. distance
   d. chemical composition
   e. temperature.

2. A star's color index is a measure of temperature because
   a. the luminosity is related to the surface area and hence to the radius
   b. the wavelength of maximum emission shifts with temperature
   c. the Doppler shift may cause red or blue shifts.

3. The method of stellar parallax is the fundamental method for determining distances to other stars. What basic fact makes the measurement of parallax possible?
   a. the Earth rotates once a day
   b. the Earth revolves around the Sun once a year
   c. stars have proper motions
   d. some stars pulsate.

4. The proper motion of a star depends on
   a. the velocity of the star across the line of sight
   b. the distance of the star
   c. the velocity of a star along the line of sight
   d. a and b
   e. b and c.

5. Variable stars vary mainly in
   a. spectral appearance
   b. magnitude
   c. direction
   d. radial velocity.

6. The temperature of a star determines its
   a. color
   b. radius
   c. mass
   d. luminosity.

7. The H–R diagram is
   a. a plot of stellar temperature versus apparent magnitude for a sample of stars
   b. a plot of stellar mass versus luminosity
   c. a plot of temperature versus absolute magnitude for a sample of stars
   d. a plot of apparent magnitude versus luminosity.

8. The absence of hydrogen lines in the spectra of type K and M stars indicates
   a. there is no hydrogen in the outer atmospheres of these stars
   b. the temperature is not high enough to excite hydrogen atoms to the level required for these lines to appear
   c. these stars are very old and have entered a period of helium burning
   d. all the hydrogen in these stars is ionized.

9. To determine the sum of the masses of a binary star, we need to measure the component
   a. temperatures and periods
   b. distance from us and size of their orbits
   c. periods and size of their orbits
   d. distance from us and the orbital velocities.

10. The absolute magnitude of a star is
    a. the magnitude of the star as seen from the Earth
    b. the magnitude a star would have at a distance of 10 parsecs
    c. a measure of the star's diameter at 10 parsecs
    d. the rank of a star in its constellation.

Answers to Selected Review Questions from the Text

1. The techniques for measuring astronomical positions used by the ancient Greeks consisted basically of using pointing devices and sighting by eye. The most contemporary of the ancient Greeks achieved accuracies a bit better than one degree; perhaps as good as 20 or 30 arcminutes. Modern techniques involve the use of telescopes capable of very accurate pointing, and the use of photographs on which careful measurements of positions can be made; the accuracies achievable by these techniques today are as good as 1/100th of an arcsecond. New techniques involving interferometry are thought capable of producing accuracies as good as 1/1000th of an arcsecond or perhaps even better.

2. Stellar parallax played essentially a negative role in the development of the heliocentric theory. Several of the ancient astronomers realized there should be a parallax motion if the Earth really orbited the Sun, yet the observational techniques of the time were incapable of detecting this motion. To these ancients, the distances to the stars implied if the parallax were below their detection threshold seemed ridiculously large. These arguments led ancients such as Hipparchus, and even some relatively recent astronomers such as Tycho Brahe (late 1500's), to conclude that the Earth must not be orbiting the Sun.

3. Recall that 1 pc = 3.26 light-years:

$$\text{nearest star}: \ 1.33 \text{ pc} \times 3.26 \ \frac{\text{ly}}{\text{pc}} = 4.34 \text{ light-years}$$

$$\text{galactic center}: \ 8000 \text{ pc} \times 3.26 \ \frac{\text{ly}}{\text{pc}} = 26,000 \text{ light-years}$$

$$\text{LMC}: \ 46,000 \text{ pc} \times 3.26 \ \frac{\text{ly}}{\text{pc}} = 150,000 \text{ light-years}$$

$$\text{Andromeda galaxy}: \ 700,000 \text{ pc} \times 3.26 \ \frac{\text{ly}}{\text{pc}} = 2.3 \text{ million light-years}$$

4. If the parallax angle for an object can be measured, then the distance to that object can be determined from d = 1/p , where d is the distance in *parsecs*, and p is the parallax angle in *arcseconds*.
   (a) A star with a parallax angle of 0.25 arcseconds is located at a distance of 4 pc.
   (b) A star with a parallax angle of 0.04 arcseconds is located at a distance of 25 pc.

(c) A star with a parallax angle of 0.005 arcseconds is located at a distance of 200 pc.

5. The magnitude scale is defined so that a difference of 5 magnitudes corresponds to a factor of 100 in brightness; thus, each magnitude represents a factor of about 2.5 in brightness. Further, the scale is "backwards" so that higher (more positive) numbers are assigned to dimmer objects. A 12^m star therefore is about 2.5, or about 40%, fainter than an 11^m star. A 4^m star is about 2.5 times as bright as a 5^m star.

10. The absolute magnitude of a source is the apparent magnitude that the source would have *if* it were located at a distance of 10 pc. The apparent (m) and absolute (M) magnitudes of a source are related to its distance (d) by (see Appendix 10):

$$ m - M = 5 \log \frac{d}{10 \text{ pc}} \quad \Longleftrightarrow \quad d = 10^{1 + \frac{m-M}{5}} \text{ pc} . $$

(a) For a star 1000 pc distant with m = 12.3 we find M = 2.3.
(b) For a star 10,000 pc distant with m = 12.3 we find M = −2.7.
(c) A star at a distance of 10 pc has the same apparent and absolute magnitudes.
(d) For a star 1 pc distant with m = 2.1 we find M = 8.1 (in fact, there are no stars this close to the Sun).

12. A G2 star that lies above the main sequence on the H-R diagram has a greater luminosity than a G2 star that lies on the main sequence. Since both stars have the same spectral type, they must both have the same surface temperature. The only way a star can be more luminous than another star with the same temperature is by having more surface area. Therefore, we conclude that the star above the main sequence must be a giant or supergiant.

15. Newton's modification of Kepler's third law states:

$$ p^2 = \frac{4\pi^2}{G(M_1 + M_2)} a^3 . $$

If the orbital period (p) is measured in years, the semimajor axis (a) measured in AU, and the masses measured in solar units, then the law can be rewritten:

$$ p^2 = \frac{a^3}{(M_1 + M_2)} . $$

Substituting 1 year for the period, and 2 AU for the semimajor axis, we find that the system mass is $8M_\odot$.

The center of mass of the system is one-fourth of the 2 AU from the more massive component (star A) to the less massive (star B); thus, star A must therefore be three times as massive as star B: $\frac{1}{2}M_A = \frac{3}{2}M_B$. Along with $M_A + M_B = 8M_\odot$ we solve and find that $M_A = 6M_\odot$ and $M_B = 2M_\odot$.

Answers to Self Test

1–e, 2–b, 3–b, 4–d, 5–b, 6–a, 7–c, 8–b, 9–c, 10–b

15
Stellar Structure: What Makes a Star Run?

A star is a spherical ball of hot gas whose interior structure is kept in *hydrostatic equilibrium* by the balance of the inward force of gravity and the outward pressure exerted by hot gas and radiation. The visible surface of any star is called its photosphere; cool stars generally have chromospheres or coronae above the photosphere, while hot stars have strong stellar winds.

The most fundamental parameter governing the properties of a star is its mass. Indeed, a certain minimum mass (about 0.08 $M_\odot$) is required in order for the gravitational attraction to cause sufficient interior temperatures and densities to sustain nuclear fusion. (For example, Jupiter is not a star even though it has the same composition as the Sun because its interior is not hot and dense enough to sustain fusion.) Stars on the main sequence follow a *mass–luminosity* relation, with more massive stars being much more luminous. Because these massive stars are so luminous, their lifetimes actually are shorter than the lifetimes of smaller stars. Thus, we have the rule that the more massive the star, the shorter it will live. The most massive stars live only a few million years, while the least massive stars have life expectancies much greater than the current age of the universe.

Besides mass, the next most important parameter governing the structure of a star is its chemical composition, particularly in its interior. Main sequence stars are composed mostly of hydrogen, and produce energy by the fusion of hydrogen into helium. The sequence of steps followed in the hydrogen fusion reaction varies with the stellar mass; small, cool stars burn by the proton-proton chain (as does the Sun) while massive stars use a different sequence of steps called the *CNO cycle*. Note that both the proton-proton chain and the CNO cycle produce the same net result: the conversion of hydrogen into helium. Because hydrogen is being fused into helium, the interiors of stars become enriched in helium over time. This change in chemical composition will cause major adjustment in the interior structure of the star, and the star will leave the main sequence moving towards the giant or supergiant regions on the H–R diagram. Once a star exhausts its capacity for hydrogen fusion, it can begin to fuse helium into carbon by a sequence of steps called the *triple-alpha reaction*. Some massive stars can continue to fuse heavier and heavier elements as each successive fuel is exhausted until they may convert their cores to elements as heavy as iron.

Energy generated in the core of a star must slowly leak out towards the surface. The method by which the energy leaks out depends on how quickly the temperature decreases with distance from the center. If the temperature gradient, or drop–off with distance, is very steep then energy is transported by convection, whereas a less steep gradient will allow radiative transport of energy. Most stars have regions of each type of transport. Cool stars generally have radiative transport deep in their interiors, and convective transport closer to the surface; this convection is believed to be important in establishing chromospheres and coronae. The situation reverses in hot stars, where the cores are convective and the outer layers radiative.

Much of this chapter concerns properties of stars, such as their interior structures, which cannot be directly observed. How, then, can we speak with confidence about the center of a star? The answer lies with what we call a stellar model. This is a mathematical representation of a star in which we use known physical laws, in conjunction with the observed surface features of stars, to calculate the properties which we cannot directly observe. Once such a model has been created, it is possible to test the model: the model will allow predictions of observable properties. If the observed properties match the predictions, then the model is likely to be valid; if they do not match, then modifications of the model will be necessary. Today's mathematical models involve so many computations that they require the use of powerful supercomputers.

Words and Phrases

CNO cycle

convection

hydrostatic equilibrium

photosphere

mass–luminosity relation

radiation pressure

radiative transport

stellar evolution

stellar models

stellar winds

strong nuclear force

triple–alpha reaction

Exercise: Stellar Interiors

(Note: This exercise is somewhat more difficult than most of the others in this study guide. It is included because it will give you a good understanding of stellar modelling if you complete it.)

In this chapter we discussed the interiors of stars. Since no direct observations are possible, astronomers make mathematical models of stellar interiors, which incorporate the applicable laws of physics and the necessary observations of stellar surface conditions. One important concept is that of hydrostatic equilibrium, the balance of pressure and gravity. In this exercise, we will apply this concept to the Sun and see how we can derive the pressure at its center.

Pressure is defined as the force on a specified area. To express the outward force in the Sun due to gas pressure, we multiply the pressure by the area of the surface:

$$\text{outward force} = P(4\pi R^2)$$

The inward force holding the Sun together is gravity:

$$\text{inward force} = GM_1 M_2 / R^2$$

Let us consider the two masses in this equation to be:

M_1: a shell of solar material with radius R. We can express the mass of this shell as its volume times its density. The surface area of the shell is given by $4\pi R^2$ and the thickness of the shell is r, so the volume is $4\pi R^2 r$. The density is d, so the mass is $4\pi R^2 rd$.

M_2: all the mass interior to our shell, which we can call M(R).

Now let the outward force on the shell equal the weight of the shell pulling inward and show that:

$$P = GM(R)rd / R^2$$

Use this relation to estimate the pressure at the center of the Sun. Rather than working inward through a series of shells as would be done in a genuine stellar model interior, we will use only one shell. In this case both r and R equal the solar radius. If you use units of centimeters and grams then the constant $G = 6.67 \times 10^{-8}$. You will also need the average density of the Sun, which you can compute from the mass and volume of the Sun.

Your answer for the central pressure will be in units of dynes/cm². This may be more meaningful if you convert it to pounds/inch² using the relation 1 dyne/cm² $= 1.45 \times 10^{-5}$ lb/in². How does the pressure inside the Sun compare with sea–level atmospheric pressure on the Earth? How does it compare with the atmospheric pressure at the surface of Venus?

Self Test

1. Which of the following type of star puts out the most energy per unit area of its surface?
   a. G star
   b. B star
   c. F star
   d. A star
   e. M star.

2. Which of the following statements about massive stars is always (that is, regardless of whether the stars are on the main sequence) true?
   a. a massive star always is brighter than a less massive star
   b. a massive star always has a shorter lifetime than a less massive star
   c. a massive star always has a different chemical composition than a less massive star
   d. all of the above.

3. The reaction which changes helium into carbon is called the
   a. CNO cycle
   b. helium flash
   c. triple–alpha reaction
   d. proton–proton cycle.

4. Mass loss as the result of stellar winds
   a. is always so small that it cannot significantly change the mass of a star
   b. is large enough that it significantly changes the mass of all stars over their lifetimes (by a factor of two or more)
   c. is insignificant for hot stars and supergiant stars, but significant for small, cool stars
   d. is insignificant for small, cool stars but can be quite significant for hot stars and supergiant stars.

5. The relation between the mass of a star and its lifetime is
   a. a direct one because the more massive a star, the more fuel it has to burn
   b. an inverse one because more massive stars consume their fuel much faster than low–mass stars
   c. the lifetime is proportional to the square of the mass because of the inverse square relation
   d. an inverse proportion because nuclear reactions are less efficient at such high temperatures.

6. Which of the following types of stars is likely to have stellar winds with the highest velocity?
   a. O star
   b. G star
   c. M star
   d. supergiant stars.

7. Which of the following types of energy transport generally is not important in the interiors of stars?
   a. convection
   b. radiation
   c. conduction
   d. all of the above are important.

8. A star in hydrostatic equilibrium is in balance between the forces of
   a. gravity and mass
   b. gravity and pressure
   c. pressure and temperature
   d. temperature and nuclear reactions.

9. As a star ages, the chemical makeup of its interior changes. Which of the following occurs while the star is on the main sequence?
   a. helium is replaced by hydrogen
   b. hydrogen is replaced by helium
   c. hydrogen is replaced by carbon
   d. carbon is replaced by iron.

10. Most of what we know about the interior structures of stars has been learned by
    a. direct observation of the interiors using X–ray telescopes
    b. constructing and testing mathematical models
    c. recording seismic waves on the stellar surface
    d. studying the shapes of spectral lines formed in the photospheres.

Answers to Selected Review Questions from the Text

1. We know that a star must be in a state of balance between gravity and pressure because if it were not, it would either be contracting if gravity were stronger than pressure or expanding if pressure were overwhelming gravity.

4. For a star of $15M_\odot$ and $10,000L_\odot$, the lifetime is $15/10,000$ of the solar lifetime, or about 15 million years. For a star of $0.2M_\odot$ and $0.008L_\odot$, we find a lifetime of 250 billion years. (Note: this assumes that the same fraction of the total mass is available for fusion in all stars. Massive stars generally have convective cores, and consequently can burn a greater fraction of their hydrogen. Their lifetimes therefore are somewhat longer than the above formula predicts.)

7. As a star undergoes successive stages of nuclear burning, each subsequent stage involves heavier nuclei than the previous. Because heavier elements contain greater numbers of protons, their electric charge is greater and hence the electric repulsive force between nuclei is greater. In order to overcome the greater electrical repulsion, the speed of collision between nuclei must be larger with each subsequent stage of nuclear burning. Hence, higher and higher temperatures are needed to create reactions involving heavier and heavier nuclei.

9. A mass loss rate of $1M_\odot$ every 500,000 years is equivalent to $10M_\odot$ each 5 million years. If the star has an initial mass of $20M_\odot$, it would lose half of its original mass in 5 million years.

10. The escape velocity from the surface of any object is: $v_e = \sqrt{(2GM)/R}$, where R is the object radius, and M is the mass. The surface gravity is given by: $g = (GM)/R^2$. With a radius 200 times its present radius, the Sun's escape velocity will decrease by a factor of $\sqrt{200}$, or about 14. The surface gravity will decrease by a factor of 200^2, or about 40,000. The low escape velocity from the envelope of a red supergiant allows relatively cool gas in the outer atmosphere to escape. Thus, we often observe cool, slow winds from red supergiant stars.

Answers to Self Test

1–b, 2–b, 3–c, 4–d, 5–b, 6–a, 7–c, 8–b, 9–b, 10–b

16
Life Stories of Stars

Most astronomical events, like the evolution of stars, take place over such long periods of time that we have no hope of observing them directly. Nevertheless, by studying many stars we have been able to piece together a detailed understanding of the process of stellar evolution. Of particular importance are studies of star clusters because we can assume that all of the stars in the cluster are located at about the same distance from us, are of about the same age, and began their lives with the same chemical composition. Thus, differences we find among the stars must be the result of differences in their evolution. Three main types of clusters are identified: open, or galactic, clusters which contain typically a few hundred stars and are found in the disk of the galaxy; OB associations, which are loose associations with many stars of spectral type O or B; and globular clusters, spherical groups of hundreds of thousands of very old stars usually found outside the plane of the galaxy.

In studying a cluster we can plot a color–magnitude diagram. This is very similar to a standard H–R diagram, except the vertical axis plots apparent, rather than absolute, brightness. By comparing the cluster main sequence to the standard H–R diagram main sequence we can determine the distance to the cluster; this is the method of main sequence fitting. Since stars higher on the main sequence have larger masses and thus shorter lifetimes, we can determine the age of the cluster by finding the main sequence turn–off point; stars above this point have already ended their hydrogen burning lifetimes, so the lifetime of stars at the turn–off point is the age of the cluster.

It is believed that all stars form in clusters from the collapse of interstellar clouds. The collapse must be initiated by some kind of trigger but, once begun, gravity will ensure that the collapse continues. Fragments form in the collapsing cloud, each to become a single (or multiple) star system. As the core of the collapsing fragment heats up, it becomes a *protostar*; as the protostar continues to shrink it may go through a *T Tauri* phase with a strong stellar wind. Eventually, the core heats to the point where nuclear fusion begins. The star has then reached *zero age main sequence* and shines as a stable, main sequence star. This process is similar for stars of all masses, though the time required for star formation is much shorter for the more massive stars.

A one solar mass star, like our Sun, remains on the main sequence burning hydrogen by the proton–proton chain for about 10 billion years; during this time the core temperature gradually rises and the luminosity of the star increases slightly. When the core hydrogen is exhausted, the inert helium ash continues to shrink and heat; meanwhile, temperatures around the core rise high enough to ignite hydrogen shell burning since hydrogen still is present outside of the central core. The luminosity of the star actually increases during the shell burning phase; its surface also expands, and the star moves up the H–R diagram becoming a red giant. As it shrinks, the inert helium core becomes *degenerate*; it resists further compression because of a special property of electrons when they are squeezed close together. As a result of the properties of degenerate gases, the ignition of helium fusion occurs suddenly and violently in an event known as the *helium flash*. Following this event the star burns helium into carbon by the triple–alpha process, and its outer layers shrink back slightly so that the star moves leftward on the H–R diagram; in globular clusters we identify *horizontal branch* stars as being in this stage of their lives. In due course the star exhausts its core helium; it then enters a second red giant phase in which helium burns in a shell around the core while hydrogen burns in a second shell even further outside the center. This second red giant stage is short–lived, and the star expels its outer layers; observed examples of this process are called *planetary nebulae*. The remaining core of the star is left behind to become a *white dwarf*.

Stars more massive than the Sun evolve similarly, though with some important differences. First, the more massive stars evolve faster. Hot stars burn hydrogen on the main sequence via the CNO cycle rather than the proton–proton chain. Once they finish burning helium in their cores, their large mass may allow temperatures to go high enough to burn carbon and heavier elements. Late in its life, the core of a massive star may resemble the structure of an onion, with each layer representing the burning of a different element (heavier elements near the center). Exact predictions of the evolution of massive stars are difficult because the stars may lose mass over the course of their lifetimes through stellar winds or other ejections of matter. Nevertheless, the nuclear reactions must eventually come to an end. If the remaining mass of the core is below 1.4 solar masses the star will become a white dwarf; if it is above 1.4 solar masses, however, electron degeneracy pressure is insufficient to support the star against gravity and it cannot become a white dwarf.

It is thought that stars of mass up to about 6 or 8 $M_\odot$ can end as white dwarfs. More massive stars can proceed through their stages of nuclear burning until the core becomes iron. Because fusion of iron cannot generate energy, the collapse of the core suddenly accelerates. The star may undergo a supernova (Type II) explosion. The core left behind after the explosion may be a neutron star or, for the most massive stars, a black hole.

Although astronomers frequently witness supernovas in other galaxies, no supernova in 400 years had occurred near enough to allow detailed study until 1987. Known as Supernova 1987A, this event has revolutionized our understanding of supernova explosions. It occurred in the Large Magellanic Cloud and was visible to the naked eye for several weeks. Since the supernova still is evolving as its remnant expands into space, it will be the subject of careful study for decades to come.

Most of the chapter deals with the evolution of single stars. The evolution of stars in binary systems may be somewhat different if the stars are close enough together to allow for mass transfer. Mass transfer occurs when one member of the binary system expands towards becoming a red giant, and its expanding layers begin to flow onto its companion. Mass transfer also can occur when material blown off by a stellar wind flows to the companion. The process of mass transfer explains some otherwise paradoxical systems, such as the binary star Algol in which the more massive of the two stars is actually less evolved.

Words and Phrases

black hole	neutron star
Chandrasekhar limit	OB association
color–magnitude diagram	open cluster (galactic cluster)
degenerate gas	planetary nebula
endothermic reaction	Pleiades
globular cluster	protostar
helium flash	Ring nebula
horizontal branch stars	supernova
Hyades	supernova 1987a
main sequence fitting	T Tauri star
main sequence turn–off	white dwarf
mass transfer	Wolf–Rayet star
neutron capture reactions	zero age main sequence

Exercise: Star Clusters

The Pleiades are a bright open cluster whose brightest members (the "Seven Sisters") are visible without a telescope. In this exercise you will determine the distance to this cluster and derive some of its properties.

Below is a list of observations of some of the stars in the cluster. Each star has been studied photometrically and its magnitude, V, and color, B–V, have been measured. Plot a color–magnitude diagram for the cluster (this should be done on graph paper). Now use the method of main sequence fitting described in the text to find the distance modulus of the cluster. Use the standard H–R diagram shown in the text. You may find it helpful to redraw the standard H–R diagram to the same scale as your diagram of the Pleiades. What is the distance to the Pleiades?

Now answer the following questions about this cluster.

a. Which are the four most massive stars in your color–magnitude (i.e. H–R) diagram?

b. Which stars in your diagram do you think are the "Seven Sisters"?

Stars in the Pleiades*

Star Number	V	B-V	Star Number	V	B-V
3	8.2	.09	26	7.9	.04
5	8.1	.04	27	5.7	−.22
7	9.6	.33	28	6.4	−.20
8	8.1	.12	31	4.2	−.24
9	9.8	.41	32	10.4	.53
10	5.4	−.22	33	7.4	−.07
11	3.7	−.30	34	8.1	.21
13	10.4	.51	35	10.2	.62
15	8.6	.20	36	9.3	.34
16	5.6	−.26	39	10.5	.53
17	4.3	−.30	40	6.8	−.15
18	8.9	.31	41	8.4	.15
19	8.0	.07	42	9.4	.34
20	8.6	.21	43	6.9	−.14
21	10.1	.49	44	7.6	.05
22	7.2	.01	45	7.3	−.13
23	9.7	.43	46	7.7	.00
24	9.4	.37	47	6.8	−.10
25	3.9	−.25	56	6.9	−.03

* Data from Johnson, H. L. and Morgan, W. W. 1951, *Astrophysical Journal*, 114:522.

Self Test

1. A white dwarf cannot be more massive than
   a. the Earth
   b. the Sun
   c. about 1.4 solar masses
   d. about 10 solar masses.

2. If a cluster exhibits a main sequence turn–off at spectral type F2, it means that
   a. the cluster has no stars more massive than those of type F2
   b. stars more massive than F2 stars have not yet formed
   c. stars more massive than F2 stars have evolved off of the main sequence
   d. stars less massive than F2 stars have not yet evolved onto the main sequence.

3. When a new star like our Sun is forming, what is the event that halts the initial gravitational contraction?
   a. the onset of fusion
   b. the helium flash
   c. the core becoming degenerate
   d. the exhaustion of fuel supplies.

4. Suppose that H–R diagrams have been constructed for two clusters. Cluster No. 1 contains main sequence stars from type B through type M. Cluster No. 2 contains main sequence stars from type A through type M. Which cluster is older?
   a. No. 1
   b. No. 2
   c. impossible to tell without more data
   d. both clusters are the same age.

5. Star clusters are useful in the study of stellar evolution because we can assume that all of the cluster stars
   a. are at the same distance and were formed at about the same time
   b. are at the same distance and are of the same mass
   c. are of the same spectral type and luminosity class
   d. will all evolve in exactly the same way.

6. What is an evolutionary track?
   a. the path of a star through space
   b. a dark filamentary feature on the surface of a star
   c. a graph of temperature at various positions inside a star
   d. the sequence of positions in the H–R diagram that the star occupies as it undergoes evolutionary changes
   e. the curve traced out around a star by its planetary nebula.

7. Astronomers on the Battlestar Galactica have discovered a star that evolves in a strange fashion: its luminosity decreases while its temperature increases. In what direction does this star "move" in the HR diagram?
   a. upper right to lower left
   b. upper left to lower right
   c. straight up and down
   d. left to right
   e. lower right to upper left.

8. The probable end fate of our Sun is to become
   a. a neutron star
   b. a white dwarf
   c. a supernova remnant
   d. a black hole.

9. Supernova 1987A is important because
   a. it is the first supernova ever seen by astronomers
   b. it has allowed us our first careful study of a black hole
   c. it is the nearest supernova which we have been able to carefully study
   d. its expanding remnant might affect our climate here on Earth.

10. A star like the Sun will spend the largest fraction of its life
   a. contracting from an interstellar cloud
   b. as a main sequence star
   c. as a red giant
   d. as a planetary nebula.

Answers to Selected Review Questions from the Text

2. If two stars in a cluster have identical B−V color indices and apparent magnitudes, they must be similar in all other properties. They are the same distance from us, they have the same surface temperature, and the same luminosity; therefore, their other properties must be the same. If the pair of stars were not in the same cluster (but also had identical color indices and apparent magnitudes), we could not make any comparison of their other properties because the stars could be at very different distances. One might be a supergiant and the other a main sequence star, for example.

4. Stars that are forming go through a period where they are cool objects embedded inside interstellar clouds. Since these cool objects, called *protostars*, are intense infrared sources, infrared observations provide us the best means of actually observing star formation. Another advantage of infrared observations is that these wavelengths penetrate far into interstellar clouds so we can see into regions that are completely dark at visible wavelengths.

7. When a massive star undergoes successive nuclear reaction stages, the temperature at which each subsequent stage occurs is higher than that of the previous stage. The higher temperature causes a higher reaction rate and, hence, a higher luminosity. Because of the increase in luminosity, the lifetime in each subsequent stage must decrease because the energy is dispersed into space more quickly.

8. The average density (ρ) of an object is found by dividing its mass by its volume:

$$\text{density}: \quad \rho = \frac{M}{\frac{4}{3}\pi R^3}$$

Comparing the densities of a white dwarf and a neutron star we find:

$$\frac{\rho_{ns}}{\rho_{wd}} = \frac{M_{ns}/M_{wd}}{(R_{ns}/R_{wd})^3} = \frac{2}{(1/500)^3} = 2.5 \times 10^8 \ .$$

The neutron star is 250 million times as dense as the white dwarf.

9. We have learned that iron is the heaviest element that can be formed in stellar interiors through nuclear fusion reactions. This means, in a sense, that iron is the ultimate element that will be formed in stars. Therefore, many stars massive enough to go through successive nuclear reaction stages will end up with iron cores. These stars are also the ones most likely to explode as supernovae and return their core material to the interstellar medium to be introduced into new stars formed later. Hence, iron is relatively abundant in the universe — more abundant than the lighter weight elements that are formed during the process of building up toward iron production in a stellar core.

Answers to Self Test

1–c, 2–c, 3–a, 4–b, 5–a, 6–d, 7–a, 8–b, 9–c, 10–b

17
Stellar Remnants

Stars are born from clouds of gas in interstellar space. During their lives, stars return some of this material to space through stellar winds. In their death throes, they return more material through planetary nebulae (small stars) or supernovas (massive stars). Nevertheless, some of the stellar material is never recycled into space and remains as a stellar remnant after the death of the star.

There are three types of stellar remnants: white dwarfs, neutron stars, and black holes. The previous chapter discussed the means by which these remnants are produced; here, we deal with their properties. White dwarfs are supported against the crush of their own gravity by the balancing force created by the degenerate gas pressure of their electrons. Neutron stars are compressed to such high density that most electrons have combined with protons to form neutrons; the further collapse of a neutron star is prevented by the degeneracy pressure of the neutron gas. Neutron degeneracy pressure is much stronger than electron degeneracy pressure, which explains why a neutron star can support a much greater gravitational attraction than a white dwarf. If the crush of gravity is great enough to overcome the pressure of neutron degeneracy, then no known force in the universe can stop the further collapse of the star. Its collapse continues so that even photons of light are unable to escape its gravitational pull, and the object becomes a black hole.

White dwarfs can be detected observationally because they radiate light as a result of their high surface temperature. The spectrum of a white dwarf shows only a few, very broad emission lines; broad primarily because of the great pressure in the white dwarf atmosphere. The lines in a white dwarf spectrum also are shifted towards longer wavelengths; this *gravitational redshift* demonstrates the effect of the strong surface gravity on photons as predicted by Einstein's general theory of relativity. Because they have no means of generating more energy, single white dwarfs cool over time and eventually will become cold and dark — black dwarfs.

Many white dwarfs are found in binary systems where the companion star still is burning its nuclear fuel. In such cases, if the two stars are close enough to allow mass transfer from the companion to the white dwarf, some rather interesting phenomena can occur. As gas from the companion accretes onto the white dwarf surface, spontaneous nuclear reactions may ignite. Sometimes, if material has collected for many years, the normally dim white dwarf may become very luminous during the nuclear reactions; such an event is called a *nova*. Nova-like explosions may recur at regular intervals. If the white dwarf continues to accrete mass, it may someday exceed the 1.4 solar mass limit for white dwarfs. In that case, the white dwarf will suddenly explode in a *Type I* supernova. Supernovas of Types I and II can be distinguished observationally by their light curves.

It is believed that neutron stars can be formed only as the result of supernova explosions. Past supernovas can be detected observationally by observing the remnant gas expanding into space. In some cases these observations can be made in visible light but, more often, supernova remnants are detected in radio wavelengths where they emit by the *synchrotron* process, radiation emitted by electrons moving at high velocities in a magnetic field. The best known example of a supernova remnant is the Crab nebula; the explosion which created the Crab was witnessed on Earth in the year 1054. When a stellar core collapses to form a neutron star, angular momentum considerations demand that the neutron star be spinning rapidly. Thus, neutron stars sometimes are observed as *pulsars*, in which the neutron stars may vary in brightness many times each second, like a nearly perfect clock.

Neutron stars often can be found in supernova remnants, since it is their spinning magnetic fields which provide the energy for the observed synchrotron emission. Supernova remnants dissipate into space after a few tens of thousands of years, and the remaining neutron stars then are nearly impossible to detect. If the neutron star happens to be a member of a binary mass transfer system, however, its presence may be revealed by the X–ray emission from gas accreting onto its surface. This X–ray emission may be relatively steady, or it may occur in bursts; systems in which bursts occur are called *bursters* and are thought to be neutron star analogs of the nova process for white dwarfs in binary systems.

The accretion of material onto a neutron star in a binary system will, because of angular momentum considerations, form into a disk-like shape. Accretion disks emit radiation because the infalling material becomes very hot. If a binary system were to contain a black hole, rather than a neutron star, it still would form a similar accretion disk and radiate in the X–ray portion of the spectrum. Thus, it is possible that some of the known X–ray binary systems in fact contain a black hole, rather than a neutron star. In particular, the system known as Cygnus X–1 is thought to contain a black hole, because the unseen companion onto which mass is accreting seems to be too massive for a neutron star.

Words and Phrases

accretion disk

binary X–ray source

burster

cataclysmic variable

Crab nebula

Cygnus X–1

event horizon

gravitational redshift

light curve

nova

Occam's Razor

pulsar

recurrent nova

Schwarzschild radius

singularity

supernova remnant

synchrotron emission

Exercise: Compact Objects

In this exercise we investigate objects with very high densities. Recall that density (represented by the greek letter ρ) is defined as mass per unit volume. For a spherical object, the average density is thus given by:

$$\text{density}: \quad \rho = \frac{M}{\frac{4}{3}\pi R^3}$$

Do each of the following problems. [Note: when converting units, don't forget that there are 10^5 cm in a kilometer, which means there are 10^{10} cm^2 in 1 km^2, and 10^{15} cm^3 in 1 km^3!]

1. A typical white dwarf has a mass about that of the Sun (1 $M_\odot = 2 \times 10^{30}$ kg), and a radius about that of the Earth ($R_\oplus \approx 6400$ km).

a. Calculate the mean density of a white dwarf, in *kilograms per cubic centimeter*.

b. Compare the mass of 1 cm^3 of white dwarf material with that of some familiar object of your choice.

2. A typical neutron star has a mass of about 1.5 M$_\odot$, and a radius of 10 km.

a. Calculate the mean density of a neutron star, in *kilograms per cubic centimeter*.

b. Compare the mass of 1 cm^3 of neutron star material to the mass of Mt. Everest ($\approx 5 \times 10^{10}$ kg).

Suppose that a neutron star were to suddenly appear on the Earth. What would happen? How thick of a layer would the Earth form as it folds around the surface of the neutron star? [Hint: The volume of a spherical shell can be estimated as the surface area of the shell multiplied by its thickness. That is:

$$V_{shell} \approx 4\pi r_{shell}^2 \times (\text{thickness})$$

For this problem, we can consider the mass of the Earth (M$_\oplus \approx 6 \times 10^{24}$ kg) to be distributed in a spherical *shell* over the surface of the neutron star. We then need to calculate the thickness of a spherical shell of material on the neutron star with the same mass as the Earth. Take the radius of the shell to be the radius of the neutron star (since it will be a *thin* layer, and take the density of the material to be the mean density of the neutron star).

Self Test

1. When we observe a nova we are in fact observing a process taking place
   a. in a red star
   b. in a binary system in which both stars are supergiants
   c. on the surface of a neutron star
   d. in a close binary with one white dwarf.

2. A Type I supernova is believed to occur when
   a. a massive O star explodes at the end of its life
   b. a very young star explodes unexpectedly
   c. a white dwarf in a binary system accretes enough mass to exceed the 1.4 solar mass limit
   d. two black holes collide with each other.

3. It is possible to detect neutron stars or black holes when they are in close binary systems because
    a. the gravitational influence of the compact object causes the companion star to wobble noticeably on the sky
    b. neutron stars or black holes in binary systems constantly eject matter into space
    c. the rapid rotation of these objects in binary systems makes them visible as pulsars
    d. infalling matter from the companion star forms a hot accretion disk which radiates X–rays.

4. The expanding gas remnant from a supernova is most easily detected by
    a. observing spectral lines in visible light
    b. observing synchrotron radiation in radio light
    c. observing bursts of X–ray emission
    d. directly observing the expansion of the gas in visible light.

5. X–ray bursts are thought to occur
    a. on the surfaces of white dwarfs in close binary systems
    b. on the surfaces of neutron stars in close binary systems
    c. on the surfaces of pulsars which are not members of binary systems
    d. only during the process of supernova explosions.

6. Which of the following are rapidly rotating neutron stars?
    a. T Tauri stars
    b. Cepheids
    c. quasars
    d. planetary nebulae
    e. pulsars.

7. At the event horizon of a black hole
    a. material becomes rigidly solid so that nothing can break through
    b. time moves much faster than normal
    c. time comes to a stop
    d. anything that is in the vicinity is instantly sucked into the black hole.

8. The observation of gravitational red shifts in the spectra of compact objects verifies a prediction of
    a. Newton's laws of motion
    b. Kepler's laws of planetary motion
    c. Einstein's theory of general relativity field on light
    d. Doppler's studies of light.

9. Which of the following is believed to include a black hole?
    a. Cygnus X–1
    b. all O type stars
    c. the Pleides
    d. the Orion nebula.

10. The fate of a white dwarf is
    a. to eventually collapse to become a neutron star
    b. to eject a shell of material and become a planetary nebula
    c. to explode as a nova
    d. to cool until it no longer emits any light.

Answers to Selected Review Questions from the Text

3. A *nova* is a flare up on a white dwarf that has received additional material and usually reaches a maximum absolute magnitude of roughly −8. A *supernova*, on the other hand, involves the explosion of a star and reaches an absolute magnitude some ten magnitudes (a factor of 10^4 times) brighter. The distinction between supernovae of Type I and Type II is that a Type II supernova shows lines of hydrogen in its spectrum, whereas a Type I supernova does not. There are different mechanisms that can produce this observational result, but typically it is thought that a Type II supernova involves the explosion of a massive star which still has ordinary hydrogen gas in its outer layers. A Type I supernova generally is thought to be the explosion of a white dwarf that has accreted enough new matter to exceed the white dwarf mass limit, become unstable, and exploded.

5. The rotational speed (v) of a point on the surface is the circumference of the pulsar divided by its rotation period:

$$v = \frac{2\pi(10 \text{ km})}{0.033 \text{ s}} = 1900 \text{ km s}^{-1} = 0.006c .$$

7. If an X–ray binary system is discovered, it is usually known immediately that it must involve a compact star orbiting a normal star, with mass transfer from the normal star to the compact companion. The compact star can be either a neutron star or a black hole. The principle means by which the distinction can be made is that the black hole will have more mass than a neutron star can possibly have. Therefore, distinguishing between a neutron star and a black hole requires estimating the mass of the compact star in the system; this can be done only if sufficient information on the orbital properties of the two stars can be derived.

9. A neutron star or a black hole formed from the collapse of an ordinary star is likely to rotate very rapidly because the original star likely had some rotation. Because angular momentum is conserved as a star contracts, its rotation speed increases. Therefore, a compact remnant formed from a star that was originally rotating will be rotating very rapidly.

10. The combined mass of the system can be found with Newton's modification of Kepler's third law: $p^2 = a^3/(M_1 + M_2)$. Substituting 0.01 year for the period, and 0.12 AU for the semimajor axis, we find that the system mass is about $17M_\odot$. The compact companion therefore must have a mass of about $5M_\odot$, too big for a neutron star: it likely is a black hole.

Answer to Exercise

The average density of the white dwarf is about 1800 kg per cubic centimeter; this is approximately the mass of a large car. The average density of the neutron star is about 700 billion kg per cubic centimeter, which is about ten times the mass of Mt. Everest in less than a teaspoon!

If a neutron star suddenly appeared on the Earth, the Earth would be crushed to form a thin layer about 7 millimeters thick around its surface.

Answers to Self Test

1–d, 2–c, 3–d, 4–a, 5–b, 6–e, 7–c, 8–c, 9–a, 10–d

18
Structure and Organization of the Milky Way

Because we are inside of our own galaxy, there is no way for us to observe its overall structure directly. Furthermore, extinction of visible light caused by interstellar dust makes it difficult to see great distances into the galactic plane. Nevertheless, careful observations have led to a good picture of the Milky Way galaxy during the last 50 years. Some of the observations involve studying objects outside of the galactic plane, such as globular clusters, while others involve the study of radio and other wavelengths of light which are not blocked by interstellar dust.

Our galaxy is pancake–shaped with a central bulge. The Sun is located about two–thirds of the way out from the center. From the outside, the most distinguishing characteristic of the galaxy would be the spiral arms winding out through the disk. They define the location of the brightest stars, H II regions and young stars. There are, however, many dimmer stars between the arms and in the halo of the galaxy. The central bulge and inner part of the galaxy rotate like a solid body, while parts of the disk further out rotate differentially in accord with Kepler's third law. Application of Kepler's third law leads to a determination of the minimum mass of the galaxy as roughly 10^{11} solar masses; this neglects mass in the halo of the galaxy, however, and there is some evidence that the total mass of the galaxy might be ten times this amount.

Because the galaxy is so huge, standard distance–determination techniques (such as trigonometric parallax) fail, and new methods are needed. Cepheid variables and RR Lyrae variables, which exhibit a period–luminosity relation, have been used to determine the size scale of the galaxy. These luminous stars have pulsation periods that are related to their absolute magnitudes.

The size of the galaxy and the Sun's location within it were determined from observations of RR Lyrae variables in globular clusters, which revealed that these clusters are distributed about a point far from the Sun; and from the analysis of stellar motions near the Sun which showed that stars in our vicinity are orbiting a distant galactic center.

The spiral structure of the galaxy cannot be observed in visible light because of extinction by interstellar dust. Instead, the spiral arms are revealed by study of 21-cm radiation from atomic hydrogen. 21-cm is in the radio portion of the electromagnetic spectrum, and is not affected by interstellar dust. Further, atomic hydrogen tends to concentrate in the arms and thus presents a good picture of spiral structure.

The vast space between the stars is filled with the *interstellar medium*. The interstellar medium contains about 10% of the mass of the galactic disk; most of this mass is in the form of gas (mostly hydrogen), but a small fraction is in the form of tiny dust grains. Stars viewed through interstellar dust will appear dimmer in brightness and redder in color than they otherwise would. The interstellar medium is quite varied in structure. Some parts are heated to millions of degrees, while other regions are dense and cool.

The galactic core has been revealed to be very active. Observations suggest that a compact, massive object is located at the galactic center, which may be a massive black hole.

Words and Phrases

Cepheid variables
differential rotation
emission nebula (H II region)
halo of the galaxy
interstellar dust
interstellar medium
interstellar extinction
kiloparsec

Milky Way
nucleus of the galaxy
peculiar motion
period–luminosity relation
RR Lyrae stars
solar motion
star counts
21-cm radiation

Exercise 1: Racing Through Space

In our every day lives, we often have the impression that we are just "sitting here". But, in fact, we are attached to a planet which is rotating, revolving around a star, moving through a galaxy with that star, and moving with our galaxy through the universe. Relative to some distant quasars, we are racing away at almost the speed of light. In this exercise we will figure out the speeds at which we are racing through space.

1. A dedicated tanner realizes that the best time to "catch rays" is around local noon. However, it is not noon for long at any given place on Earth, due to the Earth's rotation. Realizing that the Earth rotates from west to east, this tanner decides to take a westward cruise in his convertible (turbo-charged, works on land or sea). By doing so, he hopes to travel against the rotation of the Earth in such a way that the Sun will remain on his local meridian for a long time. Assume that he is driving along the equator, and calculate how fast he must drive to "keep up" with the Earth's rotation. Express your answer in units of *km/hr*. (Hint: The question is equivalent to calculating the rotation speed of the Earth at the equator. Recall that the circumference (C) of a circle can be calculated from C = 2πr, where r is the radius of the circle.)

2. What is the average speed of the Earth in its orbit of the Sun? Express your answer in units of *km/hr*. (Hint: Since we are interested in an average speed, we may regard the orbit of the Earth as circular with a radius of 1 AU).

3. The Sun is located approximately 30,000 light-years from the center of our galaxy, and orbits the galaxy once every 225 million years. Calculate the average speed of the Sun in its orbit of the Milky Way. Express your answer in units of *km/hr*.

Exercise 2: Galactic Structure

Although we cannot travel outside our galaxy to look down on the disk, we can construct a map or picture from the location and distance of different stars. We measure positions in the galaxy by specifying the angular distance of objects from fixed reference directions, just as we describe the positions of stars by using right ascension and declination. This time, our reference frame is the plane of the galaxy. All positions are measured in degrees around the plane, starting at the center. Thus 0° is towards the galactic center, 90° is in the direction of rotation, 180° is towards the anticenter, and 270° is away from the direction of rotation. We also need to give the angular position north or south of the galactic plane, in order to completely specify a star's position in galactic coordinates. Using galactic coordinates and the distance of various stellar groups, we will construct a picture of spiral arms in the vicinity of the Sun. As we will be working only with objects in the galactic plane, the angular measure north or south of the plane will always be zero and can be ignored.

The following table gives the angular distance from the galactic center, called the galactic longitude, and the linear distance from the Sun for a number of OB associations. Plot these, using either regular graph paper or polar coordinate graph paper if it is available, with the Sun at the center. Answer the following questions about your figure:

a. On your plot, sketch the outline of the spiral arms
b. Where is the center of the galaxy in your figure? It will probably be off the paper.
c. How do you suppose the distance of each of these OB associations was found? How many separate observations were necessary if each association has an average of 15 stars?
d. Why do you suppose that there are no associations more distant than 4.4 kpc in this list?

Galactic OB Associations*

Galactic Longitude	Distance (kpc)	Galactic Longitude	Distance (kpc)
10°	1.6	158°	0.4
11	1.7	173	1.3
12	2.4	173	3.2
16	2.2	188	1.5
17	1.7	205	0.7
18	2.0	206	0.5
23	1.0	207	1.5
60	2.0	224	1.3
72	2.3	238	1.5
75	1.8	244	2.5
78	2.3	246	4.4
78	1.2	266	1.8
80	1.8	286	2.5
88	0.8	287	3.5
97	0.6	287	3.6
100	0.8	287	2.6
103	3.5	287	2.5
108	2.1	290	2.0
110	0.9	291	3.2
112	2.6	294	2.4
116	2.5	303	2.5
118	0.8	320	4.0
120	2.9	332	4.0
122	2.5	333	1.6
129	2.9	339	2.5
134	2.3	340	1.4
136	2.2	340	3.5
143	1.0	343	1.9
146	3.3	351	2.6

* Data from Humphreys, R. 1978, *Astrophysical Journal Supplements*, **38**:309.

Self Test

1. The Sun moves through the galaxy
   a. in a totally random manner
   b. in a direction towards the galactic center
   c. on a highly elliptical orbit with a large inclination to the galactic plane
   d. in a nearly circular orbit about the center.

2. The time it takes the Sun to complete one orbit of the galaxy is about
   a. 1 year
   b. 2×10^5 years
   c. 2×10^8 years
   d. 2×10^{11} years.

3. According to modern observations, what is the approximate distance to the center of the galaxy?
   a. 1 AU
   b. 10 parsecs
   c. 10 kiloparsecs
   d. 30 kiloparsecs.

4. 21—cm radiation is emitted by
   a. bright stars
   b. neutral hydrogen in the interstellar medium
   c. ionized hydrogen in H II regions
   d. pulsars.

5. The diameter of our galaxy is approximately
   a. 100,000 light-years
   b. 10,000 light-years
   c. 1,000 light-years
   d. 100,000 AU.

6. The absolute magnitude of a Cepheid variable star can be determined by measuring
   a. its mass
   b. its orbital velocity around the center of the galaxy
   c. its period of variability
   d. the number of magnitudes difference in brightness between its brightest and dimmest.

7. The distribution of B stars in the galactic plane shows a spiral arm pattern because:
   a. such luminous objects do not live long enough to move out of the arms
   b. their great masses do not let them overcome the attraction of the arms
   c. they continue to accumulate new material from the clouds in the arms
   d. all stars are found only within spiral arms.

8. What fraction of the mass of the galactic disk is in the form of interstellar gas or dust?
   a. about 0.01%
   b. about 1%
   c. about 10%
   d. about 50%.

9. Some of the interstellar gas in our galaxy is very hot. The source of heating for this gas is thought to be:
   a. winds from supergiant stars
   b. supernova explosions
   c. the self-gravity of the gas
   d. the coronae of small stars like the Sun.

10. What effect does interstellar dust have on the observed magnitudes and colors of a star?
    a. dims the star, but leaves color unchanged
    b. makes the star appear redder but leaves brightness unchanged
    c. makes the star both dimmer and redder
    d. no effect at all.

Answers to Selected Review Questions from the Text

2. Hydrogen, the most abundant element in the galaxy and in the interstellar medium, is useful for mapping the structure of the galaxy because it tends to be concentrated in the spiral arms. The 21-centimeter line of atomic hydrogen is a particularly useful tool because the radiation is unimpeded by interstellar dust and thus can be observed at great distances.

4. A hot star might have the same red color as a cool star due to reddening by interstellar dust. It still is possible to distinguish the true nature of the two stars, however, by obtaining a spectrum of each and determining their spectral types.

5. Absorption lines in the spectrum of a star that are formed by interstellar gas will generally be much narrower than the lines formed in the star's own spectrum and they will generally reflect a lower degree of ionization than the spectrum of the star itself. So, on the basis of width of line and the degree of ionization, it is usually possible to distinguish interstellar from stellar absorption lines. Furthermore, in most cases, the radial velocity as measured from the Doppler shift will be different for the lines formed in the interstellar cloud than the lines formed in the star.

6. Interstellar obscuration makes a star 5 magnitudes fainter than it would otherwise be; the star has an apparent magnitude of +17. Thus, if the star were not obscured by dust, its apparent magnitude would be +12 (brighter than observed by 5 magnitudes). If its absolute magnitude is +2, then the distance to the star is found with the formula:

$$d = 10^{1 + \frac{m - M}{5}} \text{ pc}$$

Plugging in +12 for m and +2 for M, we find a distance of 1000 pc. If the correction for interstellar dust were not made, we would have found the distance modulus to be 15 magnitudes, instead of 10. In that case we would find a distance of 10,000 pc, a factor of 10 too large.

9. The primary form of evidence for the great age of globular clusters is the fact that their main sequence turn–off points occur very low on the H–R diagram. This indicates that even relatively low mass stars in these clusters have had sufficient time to use up their hydrogen fuel and move off the main sequence.

10. Because globular clusters are so old, they must have formed early in the history of the Galaxy at a time before a great deal of nuclear processing had occurred in stars. This might lead us to expect that the globular clusters contain relatively low abundances of the heavy elements that are formed in stellar interiors. In short, globular clusters may have formed before many generations of stellar processing had taken place to enrich the interstellar medium.

Answer to Exercise 1

1. 1,670 km/hr. Clearly, it would be impractical to drive this fast; some supersonic airplanes, however, are capable of such speeds.

2. As we orbit the Sun, we are travelling through space at an average speed of 100,000 km/hr! (Calculation works out to 107,000 km/hr.)
3. We are racing around the Milky Way galaxy at nearly one million kilometers per hour. (Calculation works out to 900,000 km/hr.)

Answers to Self Test

1–d, 2–c, 3–c, 4–b, 5–a, 6–c, 7–a, 8–c, 9–b, 10–c

19
The Formation and Evolution of the Galaxy

Studies of stars in our galaxy reveal two distinct populations of stars, along with gradations in each. Population I stars are found in the disk and spiral arms of the galaxy. They generally are young or intermediate in age, have heavy element (that is, elements besides hydrogen and helium) abundances similar to that of our Sun, and include stars of all spectral types. Population II stars generally are found in the halo of the galaxy, where they are seen in globular clusters, or in the central bulge. A few Population II stars are seen in the disk of the galaxy, but these can be identified because they follow orbits which cross through the plane of the galaxy; thus, they appear to us to have unusually high velocities compared to other nearby stars in the disk. Population II stars always are very old, and have heavy element abundances much lower (typically a factor of 100) than that of the Sun. Because they are old, they must be low mass stars which generally are red in color. Not all stars fit neatly into either population I or II, so we often speak of intermediate or extreme representatives of each class.

The explanation for differences in heavy element abundances lies with the evolution of the galaxy. It is believed that the universe began with only hydrogen and helium as chemical constituents (though some new theories suggest there might have been small amounts of other elements), and heavier elements are produced by nuclear fusion in the cores of massive stars. When massive stars die (e.g., supernova explosions), they return much of their material to the interstellar medium including any heavy elements that they have made. Thus, the abundance of heavy elements in the interstellar medium increases as more and more generations of stars pass. Population II stars, since they all are very old, formed at a time when there were few heavy elements in the interstellar medium. Population I stars belong to later generations and have higher abundances of heavy elements because the interstellar medium already had been enriched by earlier generations of stars. There also are abundance variations with location; the decreasing heavy element abundance with distance from the galactic center reflects a decreasing rate of star formation.

Spiral arms are the most distinctive features of galaxies like our own, and also among the hardest features to explain. They cannot be simple streamers of material in the galaxy because the galactic rotation would have caused them to wind–up long ago. Instead, spiral arms must be actively maintained by some process. Today, it is thought that the spiral structure is maintained by a density wave which propagates through the galactic disk, centered on the galactic nucleus. The wave is compressional, causing gas and dust to pile up in the spiral arms; this explains why star formation is enhanced in the spiral arms. As the galaxy rotates, individual stars pass through the spiral arms. Massive stars live such short lives, however, that they are already dead by the time they leave the region of their birth; this explains why the spiral arms are outlined by bright, massive stars.

The origin of the spiral density wave is a separate question. Calculations show that a density wave can be generated in a disk of material, like a galaxy, by a gravitational disturbance. In some galaxies, the barred spirals, it is thought that the presence of the bar can generate the gravitational disturbance necessary to initiate the density waves. In our own galaxy, it may be the nearby Magellanic Clouds that provides the needed disturbance. Whatever the cause, the density waves seem to be very common since virtually all disk–shaped galaxies show spiral structure (many galaxies are not disk–shaped, however, and these elliptical galaxies appear not to have density waves).

Our current evidence suggests that the Milky Way originated as a gigantic, spherical gas cloud. Star formation began very early, before the cloud collapsed into a disk. These stars which formed early are what we now see as population II, and they are concentrated in the halo because that is where they were born. The original cloud must have been rotating, because as it collapsed it was forced into a disk–shape. Because the collapse removed gas and dust from the halo, forcing all of it into the disk, no new star formation has taken place in the halo since the time at which the population II stars formed. All new star formation occurs in the disk, which is where we find population I. The spiral structure of the galaxy probably formed soon after the collapse into a disk shape. It is likely that the galactic structure will stay very much the same for billions of years to come; as stars die they recycle much material to the interstellar medium where it is available to make future generations of stars. With each generation, however, some material never is recycled and ends up in stellar remnants. Someday in the distant future, the era of star formation will come to an end.

Words and Phrases

abundance gradients

density wave theory

disk population

heavy elements

high velocity stars

Populations I and II

Self Test

1. The age of our galaxy is about
   a. 15×10^6 years
   b. 15×10^8 years
   c. 15×10^9 years
   d. 15×10^{10} years.

2. The age of our galaxy is found from
   a. the velocity of the Sun about the galactic center and the number of trips it has completed
   b. radioactive dating of meteorites
   c. the main sequence turn–off of globular clusters
   d. mathematical models of stellar evolution.

3. There are more heavy elements in the interstellar medium today than there were when the galaxy first formed because
   a. heavy elements are created by density waves in the galaxy
   b. light elements have escaped from the galaxy, leaving only the heavier elements behind
   c. heavy elements are produced by planets during periods of star formation
   d. heavy elements are produced by fusion in massive stars and added to the interstellar medium when these stars die.

4. The masses of Pop. II stars are all less than about 1 solar mass. This is because
   a. formation of massive stars requires a higher concentration of heavy elements than were available where Pop. II stars formed
   b. massive Pop. II stars have long since ended their lives
   c. mass loss through a stellar wind has reduced the massive stars to less than one solar mass
   d. wrong: Pop. II stars include significant numbers of massive stars.

5. Which of the following is not a general characteristic of population II stars?
   a. age greater than ten billion years
   b. low metal abundance
   c. many stars of spectral types B and A
   d. stars appear red in color.

6. It is known that the maintenance of spiral structure must be an active process because
   a. we have directly observed density waves passing through the material in our solar system
   b. if it were not, the spiral arms would have long since been wound up by galactic rotation
   c. we see many examples of disk–shaped galaxies without spiral arms, suggesting that they often disappear
   d. we have witnessed spiral arms actually grow and then dissipate in other galaxies.

7. Star formation in the galaxy is thought to occur
   a. primarily on the inner edge of spiral arms
   b. primarily in the halo
   c. between the spiral arms in the galaxy
   d. only during the first billion years of the galactic history.

8. The source of the gravitational disturbance that might maintain the spiral structure of our galaxy is thought to be
   a. a bar in the center of our galaxy
   b. a giant black hole in the center of our galaxy
   c. the great galaxy in Andromeda (M31)
   d. the Magellanic Clouds.

9. The original shape of the galaxy, at the time population II stars formed, was
   a. spherical
   b. cylindrical
   c. disk–shaped
   d. exactly the same as it is today.

10. No star formation is taking place today in the galactic halo because
    a. gas and dust originally found there was expelled from the galaxy by a strong wind
    b. gas and dust originally found there fell into the disk of the galaxy as it evolved
    c. the low abundances of heavy elements in population II stars prevents new stars from forming.

Answers to Selected Review Questions from the Text

2. A gradient is the variation of a property with distance. In this chapter we discuss the gradient of elemental abundances with distance from the galactic center. In other contexts, we might talk about a temperature gradient. For example, inside of a star the temperature decreases with increasing distance from the center. The rate of decrease of temperature with distance would be called the temperature gradient.

4. As we learned earlier in the text, comets have orbits that are randomly oriented around the Sun. Comets may approach the Sun in any plane and from any direction. Unlike the planets, whose orbits all lie in nearly the same plane, the comets fill a spherical volume around the Sun. In similar fashion, the high velocity stars have randomly oriented orbits about the Galaxy which may be very highly elongated just like the orbits of comets. The high velocity stars also fill a spherical volume centered on the Galaxy, and are not all concentrated in a single plane.

5. The structure of the Galaxy's spiral arms apparently is governed by a spiral density wave much as the rings of Saturn are governed by a spiral density wave centered on Saturn. The wavelength, that is the spacing between the spirals, is very different in the two situations but the basic mechanism is the same.

6. The stars in a galaxy have individual orbital speeds and periods which can be quite distinct from the rigid rotation of the spiral density wave that forms the spiral arms. Hence, a star may be traveling either slower or faster than the spiral arm pattern and therefore will not stay in the spiral arm in which it formed. As an analogy, consider a piece of wood floating on a pond. When a wave passes by, the piece of wood will bob up and down with the wave motion but will not be carried along with the wave. In similar fashion, stars follow their own orbits as dictated by the gravitational attraction between the star

and the galaxy, whereas the spiral density waves rotate around the galaxy at a fixed speed.

9. The massive stars, even though they are rare, have contributed more to the heavy element enrichment of the galaxy than the low mass stars simply because the massive stars are the ones which form heavy elements and then explode, distributing those heavy elements into the interstellar medium. The low mass stars have long lifetimes so that even those formed early in the history of the galaxy are, in many cases, still in their hydrogen burning main sequence stage. Furthermore, the cores of low mass stars will end up as white dwarfs and will never relinquish the heavy elements they may have formed to the interstellar medium.

10. Supernovae play several roles in the evolution of any galaxy. They are responsible for distributing heavy elements, formed during successive stages of nuclear fusion in massive stellar cores, into interstellar space where they can be incorporated into new generations of stars. Furthermore, supernovae *produce* the heaviest elements in the process of exploding: all elements beyond iron in the periodic table arise during the supernova explosions. In addition, the energy injected into the interstellar medium by supernovae is thought to trigger star formation. That is, the mechanical energy in the form of shock waves traveling through the interstellar medium is thought, in many cases, to cause interstellar clouds to collapse and form stars.

Answers to Self Test

1–c, 2–c, 3–d, 4–b, 5–c, 6–b, 7–a, 8–d, 9–a, 10–b

20
Galaxies Upon Galaxies

In the constellation of Andromeda there lies a faint, fuzzy patch that can be seen with the naked eye on a clear moonless night. Sometimes called the Andromeda nebula (also known as M31), it typifies the many such objects (all the others are fainter) noted by astronomers beginning in the nineteenth century. the nature of these nebulae long was unclear, with some people believing them to be gas clouds within our own galaxy while others argued that they were separate "island universes". The question finally was resolved in 1924 when Edwin Hubble, using a new large telescope, succeeded in identifying individual Cepheid variable stars in the Andromeda nebula. He thus was able to prove conclusively that the nebula is located far outside our own Milky Way by comparing the apparent magnitudes of the Cepheids with their presumed absolute magnitudes based on the period–luminosity relation. (Note: in fact, the distinction between Cepheids of type I and II was not yet known; as a result, Hubble's initial estimate of the distance to the Andromeda galaxy was somewhat low.) Today we realize that the Andromeda nebula, like the many other "spiral nebulae" as they were known, is a full–fledged galaxy analogous to the Milky Way.

Hubble also developed our system for classifying galaxies according to their shapes. Elliptical galaxies are subdivided according to their degree of flattening. Spiral galaxies are divided into those with a central bar (barred spirals) and those without, and subdivided according to the tightness with which their spiral arms are wound. S0 galaxies are an intermediate class, with disk shapes but no spiral arms. Misfits to either class are known as Irregular galaxies, but even these show some patterns which allow them to be divided into Irregulars of Type I (some hint of spiral structure) and Type II.

Because galaxies lie so far away, we must use methods of *standard candles* to measure their distances. Only very bright stars can be distinguished individually in even nearby galaxies; because Cepheid variables are bright we can use their period–luminosity relation to help us measure distances to nearby galaxies. As one looks further, however, even bright Cepheids cannot be seen. Here, we must use other standard candles such as supernovae, the brightest star in a galaxy, H II regions, or even the galaxies themselves. A relatively new method of measuring galactic distances is that of Tulley–Fisher, in which the galactic rotation velocity is estimated based on studies of its 21–cm radiation. The further we look, the more uncertainty comes into play. Thus, our estimates of distances to the most remote galaxies may be uncertain by as much as a factor of two.

The masses of nearby galaxies can be estimated by applying Kepler's third law to the orbital motions of stars or gas clouds in the outer portions. This technique suffers when the galaxy is too far distant to see individual stars, however. For more distant spiral galaxies, mass can be estimated by constructing a plot of orbital velocity versus distance from the galactic center (a *rotation curve*); the orbital velocities can be inferred from Doppler studies of the 21–cm line. In elliptical galaxies mass is estimated by measuring the *velocity dispersion*, the range of orbital speeds seen in the galaxy. In the rare cases where pairs of galaxies are seen to be orbiting each other, direct application of Kepler's third law allows us to determine the combined mass of the two galaxies. In such cases it is found that the mass is much greater than would be found by studying only internal motions within the individual galaxies, suggesting that these galaxies have massive halos. Similarly, studies of the masses of clusters of galaxies find much larger masses when the dynamics of the clusters are studied than when individual galactic masses are added together. It may be that much of the mass in the universe is hidden in galactic halos or between galaxies in clusters, taking a form which thus far has eluded our direct detection.

The mass–to–light ratio of a galaxy compares its total mass to its total luminosity. In all cases, it is found that the mass of galaxies is dominated by stars dimmer than our own Sun. We also find that elliptical galaxies have higher M/L ratios than spiral galaxies, indicating that they have relatively fewer bright, massive stars. Further, elliptical galaxies tend to be devoid of gas and dust, suggesting that they are made up mostly of population II stars like we find in the globular clusters of our own galaxy. Thus, the form of a galaxy — spiral or elliptical — seems to be related to the question of whether star formation is an ongoing process. It is thought that elliptical galaxies arise when rapid star formation early in the galactic history depletes gas and dust before a disk can form; if gas and dust remains to fall into a disk then the galaxy will be a spiral. In some cases, collisions among galaxies may strip the gas and dust out of a spiral galaxy leading it to become an elliptical.

Galaxies are found to be grouped into clusters. Some rich clusters may contain thousands of galaxies, while others contain only a few. Few spiral galaxies are found near the centers of rich clusters, while the intracluster gas in these regions is quite hot. This suggests that any spirals which pass near the center of rich clusters are stripped of their gas and dust by gravitational interactions with other galaxies. Rich clusters often have a giant elliptical galaxy near their center, which may be the result of galactic "cannibalism". Beyond the scale of clusters, we find that the clusters themselves are organized into superclusters. These may be the largest organized structures in the universe. Considerable debate today rages over the question of whether galactic formation is "top–down", in which superclusters formed early in the universe and then fragmented into clusters and individual galaxies; or "bottom–up", in which smaller clumps of gas or stars aggregate to form galaxies and then clusters and superclusters.

Our Milky Way and the Andromeda galaxy are the two major members of a small cluster of galaxies known as the Local Group. About 30 galaxies, most of them dwarf ellipticals, are arranged rather haphazardly in a region of space about 800 kpc across. The nearest neighboring cluster is about 1100 kpc away. Two small, Type I irregular galaxies are satellites of the Milky Way; these are known as the Large and Small Magellanic Clouds. Stars in the Magellanic Clouds have a lower fraction of heavy elements than do stars in our galaxy, and therefore provide an important laboratory for many of our ideas about stellar evolution. Equally important, these galaxies allow us to calibrate standard candle methods for finding distances to more remote galaxies.

Words and Phrases

Andromeda galaxy	Magellanic Clouds
barred spirals	mass–to–light ratio
clusters of galaxies	rich cluster
dwarf ellipticals	rotation curve
elliptical galaxy	spiral galaxies
Hubble	standard candle
intracluster gas	superclusters
irregular galaxy	velocity dispersion
Local Group	

Exercise: Large Numbers

Astronomy is a science of large numbers, especially in discussions of extragalactic objects. Distances measured in megaparsecs are difficult to comprehend, but so are the numbers associated with the national debt, or a major corporate budget. Sometimes these numbers are easier to grasp if they are reduced to something more meaningful. The names given to large numbers are often confusing: a thousand is 10^3, a million is 10^6, a billion is 10^9 and a trillion is 10^{12}, at least in the United States. The term zillion has no mathematical meaning. In most other countries the term trillion applies to 10^{18}, which in the U.S. is a quintillion. The use of scientific notation avoids the confusion caused by these terms. Remember the rules for working in scientific notation.

Consider some of the following large numbers. When a fast–food chain claims that it has sold 50 billion hamburgers, what does this mean if we assume every person in the U.S. has bought equal numbers of these hamburgers? There are about 250 million people, so

$$\frac{50 \times 10^9 \text{ hamburgers}}{25 \times 10^7 \text{ people}} = 200 \text{ hamburgers per person.}$$

Try the following. The Sun is 1.5×10^8 km from the Earth. If we travelled at 1000 km/hr (on a commercial jet), how long would it take to reach the Sun? How long does it take light, travelling at 3×10^5 km/sec, to reach us from the Sun? The nearest star is about 2×10^5 times as far as the Sun. How long would it take to reach it by jet? At the speed of light?

The large Magellanic Cloud, a satellite galaxy to our Milky Way, is 55 kpc in distance. Remember that a parsec is 3.26 light years. How many years ago did light reaching us now leave the Large Magellanic Cloud? The Andromeda galaxy is 570 kpc away. For how long has light been travelling to reach us? (Incidentally, this is the most distant object that you can see without a telescope.) Finally, consider a cluster of galaxies in the constellation Ursa Major, at a distance of 280 Mpc. How long ago did the light now reaching Earth leave that cluster?

Self Test

1. Whether a galaxy is spiral or elliptical is thought to be determined (at least in some cases) by
    a. the age of the galaxy
    b. the total mass of the galaxy
    c. the abundance of helium in the galaxy
    d. the rate of star formation early in the galaxy's history.

2. Which of the following is not a general characteristic of most elliptical galaxies?
    a. they contain mostly population II stars
    b. they contain little or no gas and dust
    c. they are much more massive than spiral galaxies
    d. they have many stars which are red in color.

3. Hubble proved that the Andromeda galaxy lies far from the Milky Way by measuring its distance through
    a. use of Cepheid variables as standard candles
    b. trigonometric parallax
    c. main sequence fitting
    d. use of a supernova as a standard candle.

4. Suppose we find a pair of galaxies which orbit each other. Which of the following methods of estimating the masses of the galaxies is likely to yield the greatest mass?
    a. applying Kepler's third law to the orbits of stars in each individual galaxy
    b. constructing rotation curves for each of the galaxies individually
    c. applying Kepler's third law to the orbital motion of the two galaxies around each other
    d. all of the above methods will yield the same results for the masses of the galaxies.

5. Which of the following is not a method of determining distances to galaxies?
    a. Cepheid period–luminosity relation
    b. measurements of annual parallax motions of galaxies
    c. assuming that all supernovae reach the same absolute magnitude at maximum brightness
    d. assuming that the brightest galaxy in any cluster is always of approximately the same intrinsic brightness.

6. The Magellanic clouds are
   a. irregular galaxies
   b. H II regions
   c. spiral arms in the Milky Way
   d. star formation regions located just a few light–years from our solar system.

7. The Andromeda galaxy and the Milky Way galaxy are both members of
   a. the Pleiades
   b. the Coma cluster
   c. a class known as elliptical galaxies
   d. the Local Group.

8. Why can we not be sure we have discovered all the members of the Local Group?
   a. our views in some directions are obscured by the gas and dust of our own Milky Way
   b. some member galaxies may be too distant for us to detect
   c. it is impossible to estimate the distances to some small galaxies so we cannot determine whether they are members of the Local Group or some other cluster
   d. observations must be carried out from space.

9. Which statement best represents our current understanding of the formation of galaxies, clusters, and superclusters?
   a. Superclusters formed first, then fragmented into the smaller units
   b. individual galaxies formed first, then aggregated into clusters and superclusters
   c. galaxies are the only real entities, and clusters and superclusters are illusions we see because of our inability to measure exact distances in the universe
   d. both (a) and (b) are considered possible, and current evidence does not allow us to definitively choose one over the other.

10. Evidence that much of the mass in the universe may be in a form which we cannot directly detect includes
    a. studies of the motions of galaxies in a cluster
    b. the rotation curves of spiral galaxies
    c. application of Kepler's third law to binary galaxies
    d. all of the above
    e. none of the above.

Answers to Selected Review Questions From the Text

2. Because galaxies of all types seem to contain comparably old objects, it is believed that all galaxies are roughly the same age. Consequently, it is believed that both spiral and elliptical galaxies were formed at about the same time, 10 to 20 billion years ago. If this is true, the Hubble tuning fork diagram cannot be an evolutionary sequence.

4. Using the relation between apparent magnitude, absolute magnitude, and distance we find the distance to the supernova is:

$$d = 10^{1+\frac{m-M}{5}} \ pc = 10^{1+\frac{24-(-19)}{5}} = 4 \times 10^9 \ pc \ .$$

7. The relative numbers of galaxies of different types in a rich cluster is influenced by processes that can convert a galaxy from one type to another. In particular, spiral galaxies in a rich cluster can be converted to elliptical galaxies by the drag force applied to a galaxy as it passes through an intracluster medium, or by near encounters between galaxies. These processes do not work well in small, sparse clusters like the Local Group; thus, the ratio of spirals to ellipticals may be different in clusters of different size and density.

8. Such a galaxy could be detected to a distance of:

$$d = 10^{1+\frac{m-M}{5}} \ pc = 10^{1+\frac{20-(-15)}{5}} = 10^8 \ pc \ .$$

100 Mpc is well beyond the Virgo cluster of galaxies located 15 Mpc away.

10. There are two general methods for estimating the mass of a cluster of galaxies. One is to estimate the masses of all the individual member galaxies and then add them up. The second method is based on the gravitational interaction of the cluster members with each other, and is much like the velocity dispersion technique for measuring the masses of elliptical galaxies; the orbital velocities of individual galaxies in a cluster are dependent on the amount of mass in the rest of the cluster. The two techniques yield very different results: the mass determined from the gravitational effects typically is some ten times greater than the mass estimated by counting up the masses of the individual galaxies. This implies that there may be a great deal of unseen matter in clusters of galaxies.

11. The hot, intracluster gas observed in rich clusters of galaxies contains heavy elements such as iron. Since iron is a heavy element known to be produced only inside of stars, and dispersed through supernovae, it is thought that the intracluster gas must be material that has been processed through stars. Therefore, the intracluster gas must have originated in galaxies where star formation and evolution can occur.

Answers to Self–Test

1–d, 2–c, 3–a, 4–c, 5–b, 6–a, 7–d, 8–a, 9–d, 10–d

21
Universal Expansion and the Cosmic Background

The discovery that the universe is expanding must rank as one of the most important and surprising achievements in human history. The discovery was announced by Hubble in 1929. Following on the work of Slipher and others he found that all galaxies outside our Local Group are moving away from us, and that the speed of recession of a galaxy is directly proportional to its distance. It is important to note that this relation arises as a natural result of a uniform expansion — study the raisin cake analogy in the text carefully (figure 21.3)! Thus, anyone located anywhere in the universe would see the same relation between galactic distances and recession speeds; the universe has no center.

Distances to remote galaxies are measured with standard candle techniques, while their radial velocities are measured by their Doppler shifts. If we make a graph of radial velocity against distance, we find a straight line where the slope indicates the rate of expansion of the universe. The straight line relation is known as the Hubble law, and the slope of the line is known as the Hubble constant. In equation form, we write the Hubble law as $v = Hd$, where v is the radial velocity of a galaxy, d is its distance (in Mpc), and H is the Hubble constant. Because our methods of measuring galactic distances have changed over time, our estimate of the numerical value of the Hubble constant is continually being refined. Today, we believe that the numerical value of the Hubble constant is between 50 and 100 km/s/Mpc.

The fact that the universe is expanding today suggests that it was smaller in the past. In fact, we can use the Hubble law to estimate when the universe was entirely concentrated at a single point. Assuming that the rate of expansion has always been the same, we find that the age of the universe is given by $1/H$, where H is the Hubble constant (note, you must convert the Mpc in the Hubble constant into units of km in order to do the calculation of age). (In reality, it is likely that the rate of expansion is not constant, since gravity has probably slowed the expansion over time. Thus, $1/H$ is actually an upper limit to the age of the universe.) Using today's estimates of the Hubble constant we find that the implied age of the universe is between 10 and 20 billion years.

Although the current expansion of the universe strongly suggests that the universe began from a single point in an event known as the Big Bang, it is by no means conclusive proof. A great deal of additional evidence in favor of the Big Bang theory has been accumulated over the past several decades, however, and the vast majority of astronomers are today convinced of the reality of the Big Bang. The most important evidence is the presence of the microwave background that pervades space. The background radiation was first detected in 1965; it corresponds to a temperature of about 3 Kelvin, and appears to be isotropic — the same in all directions. The background radiation is thought to be the remnant of the intense heat of the Big Bang; its temperature of 3 Kelvin is approximately as expected given ten billion years of cooling since the universe began.

Words and Phrases

anisotropy

big bang theory

Hubble constant

Hubble law

isotropic

3 K background radiation

Exercise: The Expanding Universe

The Hubble relation is a simple algebraic equation, but the data necessary to establish it were very difficult to obtain. In this exercise you will plot velocity versus distance for a number of clusters of galaxies. Although the relationship of velocity and distance is described as linear (in other, words, plotting one against the other produces a straight line) you will find that the actual data do not lie exactly on a single straight line. The value of H, the Hubble constant, is just the slope of the line drawn through the points in your graph. Since there is an error associated with each measurement, you must decide how the errors affect your determination of H. Errors in the measured velocities are not large. Let us assume them to be correct to ± 100 km/sec. Errors in the distance measurements, however, as determined from standard candles, can be quite large, and they increase with distance. Suppose these errors are 10 percent of the distance. Plot the data in the following table and put these error bars on the data points in your figure. Next draw what you consider to be the best straight line through the points, and determine H from the slope of this line (the slope of the line is simply the vertical coordinate at any point on it, divided by the horizontal coordinate for that point).

Now answer the following questions, using your diagram:

1. Use your value of H to calculate the age of the universe. What assumptions can you think of that are implicit in the age you have computed?

2. A cluster of galaxies in Hydra has a velocity of 60,600 km/s. How far away is it, in Mpc? In light years?

Clusters of Galaxies*

Cluster	Distance	Radial Velocity
Virgo	19 Mpc	1,150 km/sec
Pegasus	65	3,800
Pisces	66	5,000
Cancer	80	4,800
Perseus	97	5,400
Coma	113	6,700
Hercules	175	10,300
Ursa Major I	270	15,400
Leo	310	19,500
Gemini	350	23,200
Corona Borealis	350	21,600
Bootes	650	39,400
Ursa Major II	680	41,000

* Adapted from Allen, C. 1973, *Astrophysical Quantities* (London: Athlone Press 1973), p. 292

Self Test

1. The Hubble constant measures
   - a. the rate at which the universe is expanding
   - b. the distances to galaxies
   - c. the temperature of the cosmic background
   - d. the speed of light.

2. The fact that more distant galaxies are moving away from us at faster speeds than nearby galaxies indicates that
   - a. the Milky Way is near the center of the universe
   - b. the universe is expanding
   - c. the universe is hot
   - d. all of the above.

3. Which of the following is the best estimate of the age of the universe?
   - a. 10,000 years
   - b. 4.5 billion years
   - c. 15 billion years
   - d. 1 trillion years.

4. The fact that the Big Bang was hot (in temperature) is revealed by
   - a. the fact that galaxies are organized in clusters
   - b. the Hubble law
   - c. the microwave background
   - d. the fact that stellar interiors are hot.

5. Galaxies in the Local Group
   - a. follow the Hubble law
   - b. do not follow the Hubble law because the law is wrong
   - c. do not follow the Hubble law because they are gravitationally bound with our own Milky Way
   - d. do not follow the Hubble law because they formed only recently in the history of the universe.

6. Which of the following statements about the microwave background is not true?
   - a. it has a temperature of about 3 Kelvin
   - b. it is produced primarily by hot stars
   - c. it is the remnant of the radiation in the Big Bang
   - d. except for effects due to the Earth's own motion, it is virtually the same in all directions.

7. Suppose the Hubble constant is 90 km/s/Mpc. If a galaxy is receding from us at 9,000 km/sec, then its distance is about
   a. 100 Mpc
   b. 810,000 Mpc
   c. 90,000 light–years
   d. 9,000 km.

8. About 6 billion years ago, the heavy elements now found in your body most probably were
   a. being blown into space by the solar wind from our own Sun
   b. deep in the core of planet Earth
   c. inside of some unknown massive star, whose supernova released the elements into interstellar space
   d. part of the hot fireball that created the universe.

9. Although galaxies in a far away cluster are all at approximately the same distance from us, they may not all have the same redshift. The primary reason is
   a. we are unable to measure redshifts accurately so we should not expect to get clear results
   b. in addition to the recession implied by the Hubble law galaxies also have motions relative to their own cluster
   c. large galaxies may have a gravitational redshift which far outweighs the Doppler redshift
   d. there is no reason to expect any relation between the distance of a galaxy and its redshift.

10. The anisotropies which have been measured in the 3 K background radiation seem to indicate that
   a. the Milky Way galaxy is stationary relative to the background radiation
   b. the age of the universe is about 15 billion years
   c. the Local Group is moving towards some very massive source of gravitation
   d. the Earth is the center of the universe.

Answers to Selected Review Questions from the Text

2. Within a cluster, each individual galaxy will have its own orbital motion about the cluster center of mass. Thus, as viewed from within the cluster, the individual galaxies will appear to be moving with random velocities; viewed from afar, however, the entire cluster (i.e., the cluster center of mass) will be moving with the universal expansion. When we view a distant cluster of galaxies, we will observe that the average velocity of the galaxies is that expected from the expansion of the universe; the actual velocities of the individual galaxies, however, will be randomly distributed about that average.

3. The age of the universe is the reciprocal of the Hubble constant: age $= 1/H$. If the average value of H over the history of the universe is 80 km/s/Mpc, then the age of the universe is:

$$\text{age} = \frac{1}{(80 \text{ km s}^{-1} \text{ Mpc}^{-1})/(3.1 \times 10^{19} \text{ km Mpc}^{-1})}$$

$$= 3.9 \times 10^{19} \text{ s} = 12.5 \text{ billion years} .$$

If instead of using the average value for the Hubble constant over history we used the present value of the Hubble constant (assumed to be 75 km/s/Mpc in this problem), we would have derived an age of 13.3 billion years: nearly a billion years older than the true age.

4. The Hubble constant is measured by determining both the recession velocities and the distances for a large sample of galaxies. Determination of the velocities is straightforward as they can be measured quite accurately from the redshifts, but the distances are difficult to determine because measurement depends on accurate calibration of our standard candles. One of the reasons that such calibration is difficult is that when we are looking at galaxies that are very far away, we are seeing them as they were long ago; some of our standard candles, such as the brightest star in a galaxy or supernova explosions, may not have had the same luminosities long ago as they do now. Hence, the major uncertainty in determining the Hubble constant is the uncertainty in determining the distances to very far away galaxies.

6. We find the recession speeds of the galaxies from the non–relativistic Doppler formula ($v/c = \Delta\lambda/\lambda$), and their distances from the Hubble law ($d = v/H$). Assuming that $H = 75$ km/s/Mpc, we find that Galaxy A has a recession speed of 230 km/s and lies at a distance of about 3 Mpc; Galaxy B has a recession speed of 7200 km/s and its distance is about 96 Mpc; Galaxy C has a recession speed of 690 km/s and its distance is about 9 Mpc.

7. The validity of the distance measurement techniques applied to distant objects depends on the accuracy of our techniques for nearby objects. For example, our measurement of distances to the nearest stars are based on stellar parallax, which uses the Sun–Earth distance as a baseline. Thus, if new information caused us to change our idea of the Sun–Earth distance, it also would alter the distances computed for nearby stars. This, in turn, would change our calibration of the distance determination techniques based on spectroscopic parallax, and on the period–luminosity relationship of variable stars. That would alter the distances we measure to nearby galaxies which, in turn, would change the standard candles that we use to measure distances to more distant galaxies.

9. The isotropy of the background radiation is important in order to distinguish whether it is remnant radiation from the early universe or whether it might be the result of a wide range of discrete sources. If the radiation came from discrete sources, then it might have been expected that the radiation would be spotty or patchy. In contrast, the fact that the radiation is isotropic suggests that it is a characteristic of the universe as a whole.

Answers to Self Test

1–a, 2–b, 3–c, 4–c, 5–c, 6–b, 7–a, 8–c, 9–b, 10–c

22
Peculiar Galaxies, Active Nuclei, and Quasars

A number of bizarre, extragalactic objects have been discovered in recent years including radio galaxies, Seyfert galaxies, BL Lac objects, and quasars. The first three of these certainly are galaxies with unusual properties; the nature of quasars still is under debate, but evidence is mounting that they also are galaxies, in which we are seeing an active, young nucleus.

While all galaxies emit radio waves, those known as radio galaxies are unusually strong radio emitters. The radio emission is produced by the synchrotron process, or in some cases by inverse Compton scattering. In both cases, the implications are that a vast source of energy must be producing the rapidly moving electrons responsible for the radiation. The radio emission often comes from radio lobes located well outside of the visible galaxy, and many radio galaxies also show jets of material that appear to be emerging from the galactic center.

Seyfert galaxies and BL Lacertae objects are spiral and elliptical galaxies, respectively, with bright, active nuclei. The BL Lac objects, in particular, bear great resemblance to quasars although they generally have smaller redshifts. Quasars look like stars through even the largest telescopes, but their spectra show large redshifts which, if due to the Doppler effect, indicate velocities of recession as high as 92 percent of the velocity of light. There still is controversy over the cause of the quasar redshifts, but most astronomers now believe they are cosmological; that is, the large redshifts are the result of the expansion of the universe indicating that the quasars are very far away. The evidence for the cosmological interpretation of the redshifts includes the fact that some quasars now have been identified with remote clusters of galaxies, and the similarity of quasars to BL Lac objects and Seyfert galaxies.

If quasars are indeed very far away, then we are seeing them as they were in the distant past. Thus, we would be observing a much earlier epoch in the universe, and quasars would represent objects that are much younger than nearby galaxies. It may be that quasars are the nuclei of very young galaxies, and that BL Lacs and Seyferts represent an intermediate stage between quasars and normal galaxies. The nuclei of nearby galaxies, and the nucleus of our own Milky Way, also show evidence of activity, though at a much lower level than quasars. Perhaps many or all galaxies go through a quasar phase when they are young.

If we accept the hypothesis that quasars are the nuclei of very distant (and thus young) galaxies, then we must explain their extraordinary luminosity. The luminosities of quasars exceed those of the brightest normal galaxies by factors of up to 100 or 1000. Further, many quasars are variable, with periods as short as a few days. This limits the size of the light–emitting region to a few light–days at most. Their spectra are complex and show both emission and absorption lines of hydrogen and helium and often carbon, nitrogen and oxygen. The most widely accepted hypothesis at present suggests that quasars, as well as the other active galactic nuclei we have discussed, are powered by the accretion of matter into a massive black hole. This idea can explain all of the basic characteristics which we have observed in active galaxies, but the question is by no means settled.

Words and Phrases

BL Lac objects
cosmological distances
jets
giant black holes

quasars
radio galaxies
Seyfert galaxies

Exercise: Quasars at Cosmological Distances

As summarized in the text, there is good evidence that the large redshifts in quasar spectra are due to the expansion of the universe, and that these strange objects represent a very early stage in galaxy formation. To see how they compare in distance with galaxies, let us compare their derived distances with those of the galaxies we used to find the Hubble constant in the exercise in Chapter 21. The following table gives a list of quasars. Add them to the plot you constructed in the exercise in Chapter 21. You will first have to calculate the distances to the quasars in the table, using the Hubble law $v = Hd$ and the value $H = 55$ km/sec/Mpc. You will probably have to extend both axes of the diagram considerably. Use a different symbol in plotting so that galaxies are not confused with quasars.

After you have made your diagram, answer the following questions:

1. Is there any overlap in the distances of galaxies and quasars?
2. In the case of galaxies, the apparent magnitude of the tenth brightest galaxy in a cluster, a standard candle, decreases smoothly with distance. At 66 Mpc the apparent magnitude is $m = 13$, at 270 Mpc $m = 16$, and at 650 Mpc $m = 18$. Describe how the apparent magnitude of the quasars varies with distance. What interpretation can you make about this?
3. What is the absolute magnitude of a quasar at a distance of 1000 Mpc whose apparent magnitude is 15? Compare your answer with the average absolute magnitude of a giant elliptical galaxy, which is about −23.
4. How long ago did the light from the most distant quasar in the table leave its source?

Table of Quasars*

Quasar	Velocity	Apparent Magnitude
3C2	180,000 km/sec	19
3C9	240,000	18
PH957	260,000	17
3C48	77,000	16
PH1377	210,000	16
3C138	150,000	19
3C147	110,000	18
3C273	48,000	13
3C275	110,000	19
3C334	110,000	17
3C351	77,000	15

* Adapted from Allen, C. 1973, *Astrophysical Quantities* (London: Athlone Press), p. 291.

Self Test

1. In the largest telescopes, a quasar looks like
   a. a faint planet
   b. a star
   c. a dwarf galaxy
   d. a planetary nebula.

2. Radio galaxies are so named because
   a. they can only be detected in the radio region
   b. they are the only galaxies which emit radio waves
   c. their radio emission is very strong compared to that of normal galaxies
   d. none of the above.

3. BL Lac objects
   a. are galaxies with many properties of a quasar
   b. always show two jets protruding from each side
   c. have the largest known redshifts of any class of object
   d. are found at the centers of quasars.

4. The size of the emitting region of a quasar can be estimated from
    a. the broadening of its emission lines
    b. its redshift
    c. the period of its light fluctuations
    d. 21–cm radio emission.

5. The spectra of active galaxies show clear evidence of
    a. strong gravitational redshifts
    b. the presence of a giant black hole
    c. emission by the synchrotron process
    d. the presence of uranium in the galaxies.

6. Assuming their distance to be cosmological, the absolute luminosity of a quasar can be compared with
    a. a supernova
    b. our entire galaxy
    c. the brightest known galaxies
    d. the combined luminosity of tens or hundreds of bright galaxies.

7. Evidence that quasar redshifts are cosmological includes all of the following except
    a. the identification of quasars as members of clusters of galaxies
    b. their similarity to BL Lac objects
    c. their broad emission lines
    d. the detection of "fuzziness" surrounding quasars in some observations.

8. If quasar redshifts are indeed cosmological then all quasars are located at great distances. Which of the following statements would be supported by this fact?
    a. the universe is evolving over time
    b. the universe is the same today as it has always been
    c. the universe is expanding
    d. the universe is dying.

9. The absorption lines in the spectra of quasars are probably formed by
    a. supernova remnant gas in the Milky Way
    b. hot gas in the outer layers of quasars
    c. the halos of galaxies between us and the quasars
    d. the interplanetary wind.

10. The most widely accepted explanation for the large luminosities of active galaxies is
    a. multiple supernovas in the galactic core
    b. accretion into a massive black hole
    c. matter–antimatter annihilation
    d. particles flying out of a giant black hole.

Answers to Selected Review Questions From the Text

2. Astronomers suggest that massive black holes may be responsible for the vast amounts of energy emitted from the cores of radio galaxies, Seyfert galaxies, and quasars because it is known that the sources of these massive amounts of energy are always very small and very massive. The only known mechanism which is capable of producing these vast amounts of energy from a very small volume is gravitational infall to a black hole.

5. Since the redshift is large, we must use the relativistic Doppler formula (found in appendix 14):

$$z = \sqrt{\frac{c+v}{c-v}} - 1 \iff \frac{v}{c} = \frac{(z+1)^2 - 1}{(z+1)^2 + 1} .$$

Substituting $z = 2.45$, we find a recession speed of $0.84c$, or $250,000$ km/s. If the Hubble constant is $H = 75$ km/s/Mpc, then the distance to the quasar is:

$$d = \frac{v}{H} = \frac{250,000}{75} = 3,300 \text{ Mpc} = 11 \text{ billion light–years} .$$

If the apparent magnitude of the quasar is $+18$, then its absolute magnitude is:

$$M = m - 5\log\frac{d}{10} = 18 - 5\log\frac{3.3 \times 10^9}{10} = -24.6 .$$

Comparing the luminosity of the quasar to the luminosity of the Sun (absolute magnitude about $+5$) we find:

$$\frac{b_{quasar}}{b_\odot} = 10^{\frac{5-(-25)}{2.5}} = 10^{12}$$

The luminosity of the quasar is about a trillion Suns.

7. Quasar *emission* lines come from the quasar itself and have redshifts determined by the motion of the quasar. The *absorption* lines, on the other hand, are formed in the halos of intervening galaxies along the line–of–sight to the quasar. Thus, since the absorbing material is closer than the quasar, it will show a smaller redshift. If at least some emission redshifts were non–cosmological, then the absorption spectrum of a quasar showing a small (non–cosmologically produced) emission redshift could be produced by a higher (cosmologically produced) redshift in the intervening galaxy; no such situation has been observed.

8. If quasars are young galaxies, then this tells us that galaxies change radically in their overall properties during the course of their evolutions. A very young galaxy may look like a quasar to a distant observer. At intermediate ages it may, therefore, have a highly luminous core, and at much later ages it may look like an ordinary galaxy. If we wish to explore the vast distances of the universe by using far away galaxies as standard candles, then we must be aware of the possibility that galaxies very far away may have rather different luminosities than galaxies nearby. This may render very inaccurate any distances based on galaxy luminosities as standard candles.

Answers to Self–Test

1–b, 2–c, 3–a, 4–c, 5–c, 6–d, 7–c, 8–a, 9–c, 10–b

23
Cosmology: Past, Present, and Future of the Universe

We must make certain basic assumptions about the nature of our universe in order to study it as a whole. For example, we must assume that the laws of nature as we understand them on the Earth today must also apply throughout our universe both in space and in time. Modern cosmology generally begins with the additional assumptions that the universe, at least on the largest scales, is homogenous and isotropic; this is known as the *Cosmological Principle.*

To understand the large–scale structure of the universe we must invoke Einstein's general theory of relativity. One consequence of this theory is that space is curved by gravitational fields. Thus, there may be a substantial curvature near a massive object, but there also may be an overall curvature to the universe as a whole. There are essentially three possible overall structures to the universe: 1) the universe may be open, or negatively curved. In this case the universe will continue to expand forever; 2) the universe may be closed, or positively curved. In this case the universe will someday stop expanding and begin to contract; and 3) the universe may be flat, in which case the expansion eventually will come to a stop, but not reverse itself.

To answer the question of whether the universe is open, closed, or flat we must either determine the total mass density of the universe or measure the deceleration of the universal expansion. Efforts to determine the total mass density indicate that our universe is open; however, we have seen evidence that there may be substantial amounts of dark matter, which we cannot directly detect, in places such as galactic halos or clusters of galaxies. Thus, we are unable to be certain that we are properly counting all of the matter in the universe. Although it is difficult to directly measure the deceleration of the universal expansion, indirect estimates may be made based upon the abundances of elements produced in the Big Bang; studies of deuterium have proven particularly useful. Again, measurements of the deceleration point to an open universe, but they are subject to many uncertainties. A relatively new method involves the counting of galaxies at great distances in the universe; the first applications of this technique suggest that the universe may be flat.

From the known laws of physics, it is possible to reconstruct the sequence of events that unfolded following the Big Bang. During the first tiny fraction of a second the fundamental particles in nature, quarks and leptons, were formed. At this early stage the universe might also have undergone a period of *inflation*, in which it expanded very rapidly. Theories of inflation offer several advantages over conventional theories of the universal expansion; they also predict that the universe should be flat. The chemical composition of the early universe was determined during the first few minutes after the Big Bang as quarks combined to form protons and neutrons, and these in turn remained as single protons — hydrogen nuclei, or combined to form helium. Except for tiny amounts of lithium and beryllium, the early universe is thought to have been entirely hydrogen and helium. Some recent theories, however, suggest that neutron capture reactions might have formed small quantities of other elements; this would explain the observed heavy element abundances in population II stars.

As the universe continued to expand, it cooled and its density decreased. Until about 500,000 years after the Big Bang the universe was radiation dominated, as photons of light were trapped by multiple scatterings from electrons. Around 500,000 years, however, photons of light became free and the universe entered its present matter dominated state; it is photons from this time that we see today as the microwave background. The low temperature, 3 Kelvin, of the microwave background is simply the result of the cooling (i.e., redshifting) of these photons since that time. The formation of galaxies is believed to have occurred within the first few billion years as a result of "clumpiness" in the expanding universe; the origin of this clumpiness still is not known, however. Today, the universe continues to expand; its ultimate fate will depend on whether or not that expansion continues indefinitely.

Words and Phrases

closed universe
Cosmological Principle
cosmology
cosmogony
critical density
dark matter
deceleration (universal)
quarks
deuterium
flat universe

general relativity
grand unified theory (GUT)
homogeneous universe
inflationary universe
isotropic universe
leptons
matter–dominated universe
open universe
radiation–dominated universe

Exercise: The Curvature of the Universe

There are additional tests for the curvature of the universe that have not been discussed here because they are "armchair" experiments; they can be done in principle, but not in practice. This does not diminish their usefulness in helping us understand the very difficult concept of a curved universe, however.

One simple test of the curvature of the universe is to measure the sum of three angles in a triangle. As you know from plane geometry, this sum should always equal 180° (if you have forgotten this, draw a few triangles of different shape, and measure the three interior angles with a protractor; the sum will always be 180°, within the accuracy of your measurements). Now consider what happens if you draw a triangle on the surface of a sphere. You can try this, using a globe, a grapefruit, or any suitable sphere. Draw triangles of different shapes and sizes, and measure the three interior angles, then answer the following questions.

1. Is your answer different than for the plane triangles?
2. Do you get the same answer for triangles of different sizes on the sphere?
3. Suppose you could measure the angles of a triangle formed by three widely–spaced galaxies. How could you determine the curvature of the universe?
4. Why could this experiment not be done by measuring the angles of a triangle drawn from the Moon to widely–separated points on the Earth?
5. What do you suppose the result would be if you drew a triangle on a saddle–shaped surface (or any concave surface)?

Self Test

1. Which of the following is implied by the Cosmological Principle?
   a. all galaxies in the universe are the same
   b. all clusters of galaxies in the universe are the same
   c. the same laws of nature operate everywhere in the universe
   d. the density of the universe is everywhere the same.

2. If the universe is closed then
   a. it will someday stop expanding and begin to contract
   b. it will expand forever
   c. it is currently contracting
   d. we are at the center of the universe.

3. Which of the following best describes how the universe expands?
   a. it expands into an existing space
   b. it expands more rapidly at the center than at the edges
   c. the edges recede into space while galaxies are left behind
   d. the universe has no center or edges; it creates both space and time as it expands.

4. In discussions of cosmology, the "critical density" refers to
   a. the density of matter at the time of the big bang
   b. the present–day density required to close the universe
   c. the density of the universe derived by summing the masses of all visible galaxies
   d. the density of the universe at the point of maximum expansion.

5. According to the theory of general relativity
   a. space and time are intertwined
   b. time runs slower in strong gravitational fields
   c. gravitational fields curve space, so that photons of light will follow curved paths
   d. our usual concepts of space and time do not apply "outside" the universe
   e. all of the above.

6. According to modern theory, the most fundamental particles in nature are
   a. electrons, protons, and neutrons
   b. neutrinos and protons
   c. quarks and leptons
   d. quarks and GUTs.

7. According to inflationary theories of the universe
   a. the universe should be flat
   b. the universe expanded very rapidly during the first fraction of a second after the Big Bang
   c. the universe should still be expanding today
   d. all of the above.

8. The radiation that we see today as the microwave background was released
   a. at the precise moment of the Big Bang
   b. about three minutes after the Big Bang
   c. about 500,000 years after the Big Bang
   d. gradually, during the process of galaxy formation in the universe.

9. Why is it so difficult to determine the average density of the universe?
   a. there may be much "dark matter" which eludes our detection
   b. we have no way of estimating the masses of distant stars
   c. our view of most of the universe is obscured by dust
   d. since the universe is expanding we get confused.

10. The chemical constituents of the universe early in its history were
   a. pure hydrogen
   b. almost entirely hydrogen and helium
   c. mostly hydrogen and helium, but about 10% other elements
   d. mostly uranium and plutonium.

Answers to Selected Review Questions from the Text

1. Even though the universe consists of clusters and superclusters of galaxies, which represent localized concentrations of matter, we can say that the universe is homogeneous on the largest scale so long as these clusters and superclusters are uniformly distributed throughout the universe.

2. The daily anisotropy in the cosmic background radiation does not violate the assumption that the universe is isotropic. It merely shows that the Earth has its own velocity with respect to the frame of reference of the background radiation.

5. It is difficult to determine the average mass density of the universe because it is evident that there is a large quantity of mass that is invisible or, at least, very dim. For example, based on observations of the motions of galaxies in clusters, we realize that we are unable to detect all of the mass in the observed visible galaxies. Similarly, observations of the motion of interstellar clouds in our own Milky Way indicate that there is a massive, but unseen, halo. Since so much mass is evidently undetectable by our present techniques, we really do not know how much mass is present in a given volume of space.

6. We convert the number density of the interstellar medium, given as about 1 hydrogen atom for every 10 cm^3, to a mass density. Assuming the mass of a hydrogen atom is about 1.7×10^{-24} grams, the mass density is:

$$\rho = \frac{1 \text{ atom}}{10 \text{ cm}^3} \times \frac{1.7 \times 10^{-24} \text{ g}}{\text{atom}} = 1.7 \times 10^{-25} \text{ g cm}^{-3} \ .$$

According to the text, a Hubble constant of $H = 55$ km/s/Mpc implies a critical density for the universe of $\rho_{\text{crit}} = 6 \times 10^{-29}$ g/cm^3. The density of the interstellar medium is larger than the critical density by a factor of nearly 3000. Nothing is implied concerning the question of whether the universe is open or closed, however: most of the volume of the universe is *not* occupied by galaxies.

7. Even if all galaxies have very massive black holes in their cores, this would not change the dilemma addressed in question 5. The reason is that estimates of masses of galaxies in clusters that are based on the gravitational interactions of the galaxies already include any mass that is hidden in black holes inside the galaxies. Even including that mass, the universe appears not to have sufficient mass density to be closed. Therefore, if the universe is closed, the unseen mass must reside outside of the galaxies.

8. Observed deuterium is thought not to have been produced in any way other than during the early stages of the Big Bang. (Although deuterium may be produced inside stars, it also is certainly destroyed inside stars and therefore does not survive to be dispersed into space.) Helium, on the other hand, is produced inside stars and dispersed into space through stellar winds and supernovae. Thus, the amount of helium now observed in space includes not only what was produced in the Big Bang, but material that subsequently has been produced in stars. Therefore, deuterium is a better indicator of the early conditions during the Big Bang.

Answers to Self Test

1–c, 2–a, 3–d, 4–b, 5–e, 6–c, 7–d, 8–c, 9–a, 10–b

24
The Chances of Companionship

The question of whether life, and particularly intelligent life, exists elsewhere in the universe has caught the imagination of the public. The vast numbers of stars in even our own Milky Way galaxy makes the thought that we are alone almost unfathomable; yet, we have not found any clear evidence that life exists elsewhere. In recent years we have begun to search for other civilizations by monitoring radio wavelengths seeking a signal that might have been deliberately generated. Although the odds of success cannot be known in advance, the possibility that we might contact another civilization is exciting.

In order to estimate the likelihood of finding life elsewhere we must first understand life on Earth. Life on Earth is carbon–based; we refer to the chemistry of carbon which relates to living organisms as *organic* chemistry. Experiments of the type first run by Miller and Urey have demonstrated that, in the conditions prevalent on the early Earth, complex organic molecules form easily. (Indeed, we have found evidence of organic chemistry in meteorites and even in interstellar clouds — the building blocks of life are everywhere!) Although we thus know that the building blocks of life were available on the early Earth, we do not yet know how they combined to form the first living organisms. Thus, we cannot yet estimate the probability of life evolving spontaneously in Earth–like conditions. We know that life developed on Earth within the first billion years after its formation (and possibly much earlier), but complex life forms like trees or animals arose only much later. Thus, we can assume that long time periods (billion of years) are required for the evolution of an intelligent species.

Although we cannot make a solid estimate of the number of intelligent civilizations in our galaxy, the Drake equation allows us to break the question down into numerous subunits. Thus, we are able to more easily identify the sources of uncertainty in thinking about the chances that other civilizations exist. The major factors in the Drake equation are: the rate of star formation in the Milky Way, which we know fairly well; the fraction of stars that have planetary systems and the number of Earth–like planets in each system; the fraction of those planets on which life arises; the fraction of life–bearing planets on which life evolves to an intelligent and communicative form; and, finally, the average lifetime of such a civilization. The last factor is very interesting. Even if intelligent life has arisen in many places, we may not be able to find them if they always destroy themselves after just a short technological period. Our own precarious existence here in the latter part of the 20th century testifies to the danger of self–destruction by intelligent civilizations. Perhaps contact with another civilization might provide us with knowledge that could help us survive these dangerous times.

Words and Phrases

carbon–based life
DNA
Drake equation
Miller–Urey experiment

SETI
silicon–based life
technological civilization
"water hole"

Exercise: The Drake Equation

The idea behind the Drake probability equation for estimating the number of civilizations in our galaxy is that a seemingly impossible question can often be broken down into a series of questions which are easier to answer. Here we will apply the same principle to another problem, to develop a feeling for why this works.

Suppose that you are asked how many home telephones in the U.S. are in use at this moment. Let us estimate the answer. We can break it down into a series of probabilities, as follows:

let N be the number of telephones in use at any given time;
let P be the population of the U.S.;
let T be the number of telephones per person;
let F be the frequency of telephone calls per telephone; and
let L be the length of the average call.

The population of the U.S. times the number of phones per person gives the total number of phones in the U.S. Multiplying this number times the frequency of calls per telephone yields the total number of calls per day in the U.S., and multiplying that by the average length of each call (in units of days, since our factor F is in units of calls per day) gives the number of phones in use at any given moment. Thus, the value of N is simply the product of our four factors:

$$N = P\,T\,F\,L.$$

Let us now estimate the values for the individual terms in this equation. The U.S. population is approximately $P = 2.5 \times 10^8$. Suppose there is one home phone for every four people, so the number of telephones per person is $1/4$; i.e. $T = 0.25$. Let us guess that the frequency of calls per telephone is $F = 5$ per day, and that the duration of the average call is 10 minutes, or $L = 0.007$ days. Multiplying all of our factors together to calculate N yields

$$N = P\,T\,F\,L$$

$$= (2.5 \times 10^8)(0.25)(5)(0.007)$$

$$= 2 \times 10^6 \text{ phones in use at any one time.}$$

These methods can be used to solve many similar problems. Try, for example, to answer the following: how many hamburgers are being consumed in the U.S. at this moment? how many cattle must be slaughtered each day to provide this meat? how many trees must be cut down to provide newspapers each day (if we don't recycle)?

Self Test

1. The Drake equation is intended as a means of estimating
   a. the number of planets that have life of any kind
   b. the number of technological civilizations in the entire universe
   c. the number of technological civilizations that ever have existed in the Milky Way galaxy
   d. the number of technological civilizations in the Milky Way galaxy today.

2. Which of the following best describes the methods of the Miller–Urey experiment (and other similar experiments)?
   a. the chemicals present in the early atmosphere of the Earth are mixed together and supplied with energy
   b. the chemicals present in the Earth's current atmosphere are mixed together and supplied with energy
   c. ultraviolet light is shined on an empty flask
   d. the experiment involves the careful study of fossils.

3. Experiments like the Miller–Urey experiment have succeeded in producing
   a. living organisms
   b. viruses
   c. complex organic molecules
   d. silicon–based life.

4. Why are planets around other stars so difficult to detect?
   a. planets do not emit any radiation
   b. planets are so tiny and dim in comparison to stars
   c. since planets orbit around stars we don't know where to look
   d. we can't detect planets because we know that they don't exist elsewhere.

5. If the history of the Earth were compressed into one day, the length of time in which human civilization has existed (since the time of the ancient Egyptians) would be
   a. twelve hours
   b. half an hour
   c. one minute
   d. five seconds.

6. Suppose we could build a time machine and visit the Earth 3 billion years ago. When we walked out we would
   a. find no life existing anywhere on the Earth
   b. suffocate, because the atmosphere does not contain oxygen
   c. find small mammals to be the dominant creatures
   d. observe the Sun to be much closer to the Earth than it is at present.

7. Suppose that just one in one million stars has a life–bearing planet (not necessarily technological) in orbit. About how many planets would have life in the Milky Way galaxy?
    a. only one
    b. about 50
    c. about 10,000
    d. between 100,000 and 1 million.

8. Though we occasionally listen for signals from intelligent beings, we have deliberately sent only one message ourselves; it was sent in 1974 to the globular cluster M13. If anyone there is lucky enough to receive it, we can expect a reply in about
    a. the year 1996
    b. a couple of decades
    c. 50,000 years
    d. we already have received an answer.

9. The reason that carbon is so important to living organisms is that it
    a. is soluble in water
    b. can form long chains and complex molecules
    c. is the most abundant element in the universe
    d. can combine with oxygen.

10. Why would we not expect to find intelligent life on a planet with a circular orbit at 4 A.U. from an A0 main sequence star, which has 16 times the solar luminosity?
    a. the central star is too luminous
    b. the central star emits too much ultraviolet light
    c. the central star will live a comparatively short time
    d. the planet's orbit must be elliptical.

Answers to Selected Review Questions from the Text

3. If a planet were orbiting a star exactly like the Sun, but at a distance of 0.9 A.U., it would receive about 1.23 times as much energy per unit area as the Earth [from the inverse square law of light: $(1/0.9)^2 = 1.23$]. Whether or not Earth–like life forms could evolve on such a planet is not clear. The planet receives more energy than the Earth, but less than Venus. In our own solar system, a runaway greenhouse effect occurred on Venus, while CO_2 was

164

dissolved in liquid water on the Earth. We cannot say with certainty which scenario would be more likely on the hypothetical planet.

4. Many scientists expect life elsewhere to be similar in chemistry to life on Earth because the complex molecules that appear to be required to support living species can only occur with a few chemical elements; it also appears that the chemical reactions of life require a liquid medium, such as water, in which to occur. In a gas, the interactions between atoms and molecules occur too slowly, and in a solid the mobility of the atoms and molecules is too restricted to allow these important chemical reactions to take place. Furthermore, with the possible exception of silicon, carbon is the only element with the required diversity of chemical interactions to produce organic compounds capable of the complex interactions required of living forms. (A difficulty with silicon is that it readily combines with oxygen to form SiO_2.) Finally, laboratory experiments such as the Miller–Urey experiment, the discovery of organic molecules in meteorites, and observations of organic molecules in dark interstellar clouds all indicate that organic compounds are readily available in nature.

6. A planet orbiting a Population II star would have relatively low abundances of heavy elements. These heavy elements are necessary ingredients for the organic molecules that apparently gave rise to life on the Earth. Therefore, life forms on planets orbiting Population II stars might be considered less likely than life forms on planets orbiting Population I stars like the Sun. On the other hand, a planet like the Earth contains such a tiny fraction of the total abundance of heavy elements in the solar system that the difference between the abundances in the two stellar populations might be considered insignificant by comparison. In that case, it can be argued that Earth–like planets, and therefore life, might be easily formed among population II stars as well.

8. The Drake equation is: $N = R_{*}f_{p}n_{e}f_{l}f_{i}f_{c}L$. In the text, the value of N was found to be 0.1L, assuming that $f_l = 1$ (among other assumptions). If we instead assume that $f_l = 10^{-4}$, then we find $N = 10^{-5}L$. If the lifetime of a typical civilization is $L = 10^{10}$ years, then there would be 100,000 civilizations in our galaxy today. If the lifetime is only 10^5 years, then $N = 1$ (we are alone in the galaxy). [And, if the average lifetime is only 100 years, then we might be the only civilization among the nearest 1000 galaxies, a lonely thought indeed!]

Answers to Self Test

1–d, 2–a, 3–c, 4–b, 5–d, 6–b, 7–d, 8–c, 9–b, 10–c